Cambridge Studies in Biotechnology

Editors: Sir James Baddiley, N. H. Carey, J. F. Davidson,
I. J. Higgins, W. G. Potter

3   Economic aspects of biotechnology

Economic aspects of biotechnology

ANDREW J. HACKING

*Tate and Lyle Group Research and Development,
Reading*

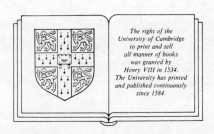

The right of the
University of Cambridge
to print and sell
all manner of books
was granted by
Henry VIII in 1534.
The University has printed
and published continuously
since 1584.

CAMBRIDGE UNIVERSITY PRESS

Cambridge
London   New York   New Rochelle
Melbourne   Sydney

Published by the Press Syndicate of the University of Cambridge
The Pitt Building, Trumpington Street, Cambridge CB2 1RP
32 East 57th Street, New York, NY 10022, USA
10 Stamford Road, Oakleigh, Melbourne 3166, Australia

First published 1986

Printed in Great Britain by the
University Press, Cambridge

British Library cataloguing in publication data
Hacking, Andrew J.
Economic aspects of biotechnology.—(Cambridge
studies in biotechnology; 3)
1. Biotechnology industries
I. Title
338.4'76606   HD9999.B442

Library of Congress cataloguing in publication data
Hacking, Andrew J.
Economic aspects of biotechnology.
(Cambridge studies in biotechnology; 3)
Bibliography: p.
Includes index.
1. Biotechnology industries. I. Title.   II. Series.
HD9999.B442H33   1985   338.4'76208   85-12729

ISBN 0 521 25893 6

Contents

Preface

The aim of this book is to relate the expanding subject of biotechnology to economic principles and general theories on industrial and technological development. It is also intended to inform academic scientists of the complexities of industrial markets, patenting, profitability and cash flow problems. A lack of knowledge of these areas was referred to in the Spinks Report on British biotechnology (p. 23) and was true of my position when I started in industrial research. Some sections of the book stem from my first experiences of costing and project appraisal. I learned that fermentation is in essence an expensive technology and many of the expenses only became apparent after a somewhat lengthy learning process. The comment made by J. B. S. Haldane in 1926: 'Why trouble to make compounds yourself when a bug will do it for you?' may be answered in part 'because it's cheaper that way'. Many biotechnological processes are not attractive when compared to the alternatives under the prevailing conditions. The problems of industrial secrecy (sometimes excessive) obscure much economic reasoning, but there are in fact sufficient data, even if most of them arise from projected costs, to permit many generalizations and conclusions. This information is however scattered and sometimes in obscure locations. To a great extent this book is a collection of facts from widely differing sources.

The book is intended for those who have a training in biotechnology and are interested in its commercial development for career purposes, to make money, or just out of curiosity. I have assumed a knowledge of biotechnology (microbiology, biochemistry, genetics) to probably at least final year undergraduate level, but have aimed to explain economics and accounting from first principles.

I have drawn the scope of biotechnology as wide as possible to encompass fermentation, enzyme technology, genetic engineering, bulk chemicals, expensive pharmaceuticals, energy and waste treatment. This may only be a passing phase. For example the General Secretary of the Society of Chemical Industry, Peter King, believes that biotechnology will split into large scale processes and expensive small volume (mostly health-care) products with little or no middle ground. By the turn of the century, with further fissions such as plant strain improvement, the term 'biotechnology' will have become obsolete. This may be so, but at present I think that there is sufficient common ground in terms of concepts, processes, raw materials and costs to link them all together.

In economic terms the intention is to discuss biotechnology both on a world and national (macroeconomic) level and in microeconomic terms of costs, investments and markets. There is more coverage of the latter and there is a bias towards bulk products as a consequence of available information. Economic analysis is strongest on existing products and processes. The newer developments are inevitably speculative and may well not develop as envisaged at present. As far as possible the approach taken is of positive rather than normative economics, that is how things are, not what they ought to be. In general I have avoided discussing future policy. In some cases values or opinions on what may be beneficial do appear, but this has been kept to a minimum.

To some extent commercial and economic terms are used which will be unfamiliar to a reader with a biotechnological background. Some are defined in the text, but explanations of others can be found in the *Penguin Dictionary of Economics*, which is very useful. Unfortunately it is not as easy to find definitions of biotechnological terminology. The *Chambers Dictionary of Science and Technology* is sometimes helpful, while various more up-to-date medical dictionaries can be used; otherwise textbooks must be consulted.

I am indebted to many people at Tate and Lyle Research and elsewhere for contributions of all kinds. I would like to mention specifically Dr Chris Bucke for much information on enzyme technology and corn wet milling, for reading the manuscript and supplying many helpful comments; Dr Peter Cheetham for data on immobilized enzymes, corn wet milling and some broad ranging ideas on biotechnology and innovation; Dr Derek Fewkes for assistance on patents; Dr Brian Hollingsworth for information on *Bacillus thuringiensis* and other microbial pesticides; Dr Chris Lawson for contributions on microbial polysaccharides in general and xanthan in particular; Brian Manston for help on the economics of anaerobic digestion; Dr George Nowacki of Miles Enzymes for interesting facts on aspects of marketing enzymes; Stuart Slocombe for sharing his extensive knowledge of capital investment; and Dr Gordon Walker for supplying data on downstream processing and immobilized enzyme processes. I would also like to thank Jane Cox and Frances De Banks for typing the manuscript, and Andrew Elliston and Jane Marsden for expert and patient information gathering.

1 The interrelationships between biotechnology and economics

The two disciplines compared

In the widest context biotechnology may be understood as referring to the commercialization of all biological processes. More precisely it has been defined as the application of scientific and engineering principles to the processing of materials by biological agents to provide goods and services (Bull, Holt and Lilly, 1982). The scientific and engineering principles are chiefly microbiology, biochemistry, genetics and biochemical and chemical engineering. It is customary not to include agriculture in this definition, although there is inevitably much overlap with agriculture, both as a supplier of raw materials and in genetic improvement of crop plants. As biology is being considered as a manufacturing technology, there must now be some analysis of costs and benefits; it must enter the sphere of economics.

Economics is most often defined as the study of the allocation of scarce resources among competing uses. Its fundamental tenets are that because productive resources are scarce, choices must be made about their use and that these choices be analysed to enable every society to make decisions on what to produce, how to produce it and for whom. In this framework biotechnological processes must be analysed both against one another and against competing technologies to determine the most beneficial route to a given commodity. Unfortunately there are aspects of both the present state of economic theory and of biotechnology which make straightforward analysis difficult. To enlarge upon this point it is necessary to outline the areas of the two disciplines which present the complications.

Biotechnology is not an industry defined by products or services such as automobiles, textiles or television. It refers to the use of microbial, animal or plant cells and enzymes to synthesize, break-down or transform materials. It is a *means of production*. Its products and services span a range of applications from food to fuel, from waste disposal to pharmaceuticals. Similarly genetic engineering is a technique to alter cells to perform functions more beneficially, that is to produce more or better products or to do so more efficiently. Genetic engineering itself cannot be subjected to economic analysis, only the processes or the improvements that it makes possible.

1

Biotechnology presents another obstacle to economic analysis in that the majority of its benefits lie in the future. It is speculative, research orientated and dependent upon a high rate of innovation at this point in time. Economists have had much more success in dealing with the consequences of technological change than with its determinants (Rosenberg, 1982). Economic theory too has tended to concentrate on issues involved in managing national demand rather than on problems of production. In particular the problems of invention and innovation and their transition into new industry are poorly understood. Similarly much biotechnological development has been undertaken with scant regard to economic issues. Technology push rather than market pull has tended to be the order of the day. A great deal of even applied research has been directed at problems which either do not justify the effort in terms of total market potential or have competing forms of technology which can do the job more efficiently. In common with some other areas such as aerospace or railways, biotechnology has engendered a fervour which has often transcended economic analysis. While having attracted funding and very favourable publicity, this phase must inevitably give way to more pragmatic considerations. Perhaps the greatest danger to advancement of the subject in the near future is a hostile overreaction to perceived failures or what in reality may be slower progress.

In common with electricity a century ago biotechnology is today a technology in search of applications. Many economic and social problems have had a biotechnological solution proposed, from precious metal extraction to a population fed by giant yeast fermentation plants. Electricity was similarly endowed with claims as diverse as sugar refining and the improvement of sexual vigour. When the early exponents of electricity such as Faraday were questioned on the utility of their discoveries they could not have been expected to predict radio, television or the laser. Similarly, with biotechnology today it is not possible to predict or even hazard a guess at all the developments that might take place. All that can be achieved is a step-wise exploration starting with the strengths and weaknesses of the technology at present.

Currently it can be stated in theory that biotechnology could have an impact across a great range of chemical substances. The list includes virtually all organic chemicals, solvents, polymers, pharmaceuticals, flavour and aroma compounds and so on. The specific products can only be identified on a case by case basis. For each chemical the relative merits of competing means of production will determine where the economic threshold for a biotechnological route will lie. Within biotechnology there will be further competition between different methods, for example immobilized cells or fermentation, batch or continuous culture.

Modern economic theory is divided into two main branches. *Microeconomics* is concerned with individual behaviour. It is the theory of small consumers, producers and markets. *Macroeconomics* is concerned with mass or aggregate behaviour. It is the theory of large scale consumer expenditure, investments and government policy. It is essential to consider biotechnology in both branches. In microeconomic terms this means looking at individual processes and plants, cost structures and management investment decisions. At a macroeconomic level it involves looking at the impact of biotechnology on national and global economics, on balance of payments questions, and issues such as energy and raw material supply.

Another complication is the uncertainty in many areas of economic theory, even at a basic level. In macroeconomic terms the workings of the modern economy are incompletely understood. There is conflict on the role of the state, arguments over monetarist and Keynesian theories and uncertainty over the best action to be taken (Godley & Cripps, 1983). This uncertainty extends to policies on biotechnology. There are conflicts of opinion on whether for example programmes to produce fuel from indigenous biomass are of net benefit, whether new industrial developments should receive government aid, whether government should be involved in industry at all. As a consequence there is a wide range of different national policies and attitudes to biotechnology.

Similarly much knowledge of industrial organization is empirical. Some of the empirical findings can be said to confirm predictions from economic theory, but the theory has rarely offered hypotheses of the accuracy with which other branches of science are used to working. Thus only limited comparisons can be drawn between biotechnology and other industrial sectors. It is not possible to say whether the same pressures that affected the computer industry will affect biotechnology or if the outcome will be the same. The development of modern businesss organization in most Western (OECD) nations has raised many controversial issues which seem to extend beyond economic theory and beyond the control of individual nations. Biotechnology then enters an arena of uncertainty where only an exploratory approach can be adopted. The decisions and their outcomes are nonetheless crucial; economic utility will ultimately shape the industry.

Limitations of cost–benefit analysis

One of the most intractable problems in the economic analysis of an industry such as biotechnology concerns how to define both the costs and benefits of the activity to society. There are several levels at which this may be examined and a number of economic models. There are also macro- and microeconomic considerations.

Firstly, economics is only one measure of the utility of an activity; while it does not necessarily apply to many areas of social activity, it is the measure industry must use. It is argued that it is the role of government to support uneconomic research when there are other reasons for doing the work. In fundamental terms industry must convert raw materials and manpower into products which meet the needs of society. In doing so it must add value and create wealth. The value of the output of the majority of industries must be greater than the resources consumed, to provide finance for such functions as social services, administration or defence, and to provide for research, development and the construction of new manufacturing plant. These tenets, it is argued, must hold – whatever political system is adopted. They are clear enough in a mature production industry such as automobile manufacture and have been used in a comprehensive text on the social and economic aspects of the chemical industry (Bradbury & Dutton, 1972). They are however less clear in new industries, particularly in the short term. Many forms of modern technology are viewed as having long lead times, but great benefits in the long term. Firms or nations which are not in at the beginning are unable to enter the race. Thus many firms which in the past have eschewed long term and unprofitable research have been prepared to make an exception of biotechnology, even to the extent of constructing large manufacturing plant which can only operate uneconomically at present, just to gain expertise. The reasons for this policy will become evident later, but it can only be a transient phase. Biotechnology will ultimately have to produce more than it consumes, but the question of measurement of costs *versus* benefits still remains.

The most commonly used method of analysis is that of cost minimization and profit maximization. It assumes that profits are the only relevant goal of a firm or manufacturing activity. Other possible goals such as social welfare or prestige are regarded as unimportant. This is the easiest way of viewing the activities of firms in the capitalist West. In terms of investment decisions it means that a firm looks for a return which exceeds that which it can obtain by other methods, either by direct investment at the prevailing rate of interest or in other manufacturing or service industries. Other factors such as risk enter into this decision, but in general it is a model which will determine a firm's behaviour and attitude towards biotechnology projects as well as elsewhere. Government may choose to make this investment favourable by selective taxation policies or import tariffs, with macroeconomic considerations in mind, but the individual manufacturer still operates within this model.

Profit is only one measure of the benefit of an activity. Other welfare factors can be cited, for example employment, lack of pollution, use of renewable resources, appropriate technology and many more. Such factors assume more importance in developing nations and in decisions on government policy. The approach used in this book in the main is on cost minimization and competition with other technologies aimed at profit maximization, with reference to other factors where appropriate. Some processes, notably in waste treatment, are not solely treated in this way, but this is made clear. The main reason for this approach is simply on the basis of available information. It is not possible to examine the policies or reasoning of Eastern Bloc nations towards biotechnology because the data are not obtainable. Similarly the policies of many developing nations, or the actual costs and benefits, are not quantifiable.

An historical outline of biotechnology

The history of a technology can prove to be an important source of information for its economic analysis. It can provide a starting point for understanding its strengths and weaknesses and its present size and importance.

The roots of modern biotechnology lie in the fermentation of foods and drinks, industries spanning almost every society and evolved over centuries by empiricism combined with adherence to tradition. Alcohol fermentations date back at least to ancient Egypt or the Biblical flood (Genesis i, 8) depending on individual beliefs. Even today the output of these industries in volume and value far exceeds new fermentation processes for the production of drugs, amino acids and industrial ethanol (Table 2.1). The essential selling points of the products, with the exception of the intoxicating effect of ethanol, are still incompletely understood. The flavour profiles and organoleptic properties of beer, wine, cheese and other products still cannot be explained in molecular terms. As a consequence the processes retain strong elements of craft and tradition, although some rationalization based on the carefully controlled conditions of the chemical fermentation industry have had notable success, for example in the American wine industry.

In economic terms, however, the craft industries and chemical fermentation are poles apart despite similarities of process and organism. A comparison of the costs of beer production and industrial alcohol illustrate this very well (Tables 5.5, 5.6 and 5.7). Excluding taxes, ethanol is sold for between $10 and $20 per litre in an impure form in dilute solution in beverages in the Western world (supermarket prices). Industrial ethanol producers must expect a price of only $0.45–1.0 per litre for 95% or absolute ethanol, which may cause many to think that they are in the wrong business. Despite their age and reliance upon traditional methods the craft fermentation industries have all shown growth which has paralleled or exceeded overall economic growth even in the most affluent societies. Because of flavour and toxicological restraints they do not have competition from other processes.

In contrast the production of bulk chemicals, such as ethanol, by fermentation has been dictated by cost, particularly in relation to the petrochemical industry. The industry developed as an outgrowth of traditional fermentation in the late nineteenth and early twentieth century to supply certain compounds to the emerging chemical and armaments industries. Early products were ethanol, glycerol, acetone and butanol, followed in the inter-war years by organic acids, notably citric, lactic, fumaric and gluconic acids. With the exception of gluconate these fermentations benefited from and ultimately were dependent upon the availability of molasses, traded increasingly on a world-wide basis as a by-product of the expanding sugar-refining industry. The technology, mostly based on stirred tank fermenters, never received sufficient impetus to develop in the way that, for example, chemistry did and the industry stagnated somewhat. It fell into a quite severe decline in the face of a greatly expanded and efficient petrochemical industry in the post-World War II years. Some processes (for example citric acid production) have survived where the product has food uses, but the industrial markets have nearly all been taken over by chemical synthesis, unless regulated by government.

The discovery of penicillin, which has now virtually passed into folklore, and the subsequent development of the antibiotic industry, has been one of the major milestones of the twentieth century and arguably one of the most beneficial discoveries of all time. Its impact on biotechnology and the public perception of biotechnology should not be underestimated. Today the antibiotic industry has sales in excess of $5 billion which ranks it alongside some of the larger products of the chemical industry. It can be said to have given biotechnology a showing in the major industrial league tables. Along with new antibiotic discoveries and commercial success there have been tremendous improvements in process development and productivity. Penicillin is now manufactured in 100 000 litre fermenters at concentrations of 50 g l^{-1} and it costs less than 1% of its cost in 1945. A less well known factor in its development has considerable significance in the economic developmemt of biotechnology today. During the Second World War the US Government imposed an excess profits tax of 85% as one measure to pay for the war effort. As many pharmaceutical companies were liable for this tax, additional research cost them effectively only 15% of its total cost, 'the 15¢ dollar'. Several of the early large antibiotic screens and the process development of penicillin were funded on this basis.

Even in the aftermath of the success of antibiotics biotechnological development was fairly slow. Enzymes were widely predicted to take the chemical industry by storm in the early 1960s, but the industry has only slowly increased to world sales of around $400 million, which is hardly large in global terms. Bulk uses are mostly confined to the starch industry and biological detergents. In detergents too growth has not been as rapid as was predicted in the 1960s, largely as a consequence of fears of allergic reactions. These have now been overcome by pelletized formations, but not before initial growth had been dented. The biggest success story of industrial enzymology has been the production of high fructose corn syrups using immobilized glucose isomerase, which is now an industry with sales of $1.4 billion. The commercial development of enzymes in general has been retarded by their expense and lability. In addition the industry still has high research and development overheads.

In Japan biotechnology has different origins from those in Europe and North America. The Japanese have traditionally produced many fermented foods, for example from soya beans. In the early part of this century they discovered some of the mould enzymes responsible for these processes, for example Takadiastase, and they developed an enzyme industry for food products from them. They also discovered that monosodium glutamate was one of the active flavour compounds in some of the fermented foods such as soy, and through strain selection and process improvements have built an industry which produces 400000 te of monosodium glutamate annually at almost bulk chemical prices. This has been followed by production of other flavour enhancing compounds such as aspartic acid and the nucleotides, plus essential amino acids such as lysine for animal feed supplements. The Japanese have invested heavily in research and have developed amino acid fermentations to a point where their nation has undisputed leadership in the field.

While sounding a death knell for many fermentation products, the low prices of oil in the 1950s and 1960s stimulated much industrial research into the production of single cell protein (SCP). Although some schemes were based on wastes or effluent streams, the greater emphasis was on petroleum products, principally methane and methanol. Many large projects were started, often by multinational oil and chemical corporations, but few ever reached a production stage. Many obstacles were encountered to this work, for example the fear that harmful hydrocarbon residues might be carried over into the product, or lack of sufficient cooling water in many otherwise ideal Middle Eastern sites. The economics of these processes were, as will be shown, always marginal because of capital and recovery costs, but they were particularly sensitive to raw material cost and they were virtually sunk in the West by the oil price rises of 1973/74. In Britain ICI and Rank

Hovis McDougall both continued with projects, the former with an animal feed product and the latter with a human food product, but both still have an uncertain future. An excellent account of the problems encountered in these processes, as well as other fermentation products, is given by Fishlock (1982). SCP plants are being constructed in Eastern bloc countries where different economic criteria apply (Chapter 5), and there may be developments in waste utilization, but otherwise interest is minimal.

Some other fermentation products have found small markets, for example *Bacillus thuringiensis* is an effective insecticide against some pests, and xanthan gum is a good emulsifier, thickener and stabilizer, but in general the widespread development of biological products has not yet occurred. Historically chemistry has remained dominant. It has supplanted fermentation in the production of ethanol and other solvents, some organic acids and several vitamins. The current interest in biotechnology is founded less on historical success than on technological expectations.

In the 1970s three events had a profound influence on the course of biotechnology. Prices of crude oil quadrupled in late 1973 and 1974, triggered at least by the Yom Kippur War. Also in 1973 Stanley Cohen and Herbert Boyer demonstrated that, with the combined use of restriction endonucleases and ligases, DNA could be cut and rejoined in new arrangements. The species barriers which had limited genetics could be overcome by *in vitro* techniques. In 1975 Georges Kohler and Cesar Milstein demonstrated the production of monoclonal antibodies from the fusion of lymphocytes and myeloma tumour cells.

The oil price rises shook Western economies to the core. Oil producing states had quite clearly gained the upper hand over industrial nations and had no interest in depleting their resources at a rapid rate to supply the rest of the world with cheap oil. The initial reaction of the Western nations was to finance developments as diverse as solar heating panels, windmills and automobiles fuelled by chicken manure. Gradual refinements and a calming of the initial hysteria gave rise to large government programmes, particularly in the US, for the development of alternative energy sources. Included in these were anaerobic digestion, fermentation ethanol production and biological degradation of lignocellulose substrates. Although the efforts in biotechnology were small by comparison to coal gasification, enhanced oil recovery or solar energy, they did provide a good stimulus for biotechnological development in general. By the late 1970s some of these programmes were looking somewhat jaded and many commercial firms were pulling out. Support decreased in 1981 with the cost-conscious approach of the Reagan Administration, as opposed to that of Carter in previous years, but some significant progress had occurred. Brazil too was committed to an extensive programme of industrial ethanol production from sugar cane. In anaerobic digestion higher reaction rates were achieved through improved reactor designs and pilot schemes were operating throughout the world. In common with fermentation ethanol however they were not yet competitive, without subsidy, with petroleum or natural gas.

The discovery of monoclonal antibodies and their applications in the detection and purification of many biological compounds has created a rapidly expanding industry comprised of small highly innovative firms competing in the medical, diagnostic and pharmaceutical fields. Many chemical, biological and radioimmuno-assays can be replaced by cheaper and more sensitive methods.

The application of *in vitro* genetic recombination has had the greatest impact, leading to the establishment of over 200 venture companies in the US with products aimed at pharmaceutical, food and agricultural markets. The most rapid and successful developments have been the bacterial production of human insulin and growth hormone, but the implications and possibilities are still difficult to appreciate fully. The immediate financial fate of many of the companies now hinges on the success of a few more advanced products, but there are signs that revenue from product sales is growing.

The basis for optimism

Atkinson & Mavituna (1983) have outlined three major factors on which the current interest in biotechnology and belief in its expansion are founded:

(1)   Biotechnology can utilize raw materials obtained from renewable resources. Cereal crops are the first contenders, but eventually celluloses and lignocelluloses could prove more important if commercially viable processes for their breakdown can be made available.

(2)   Biotechnological processes appear to have advantages over the chemical processing of vegetable materials. Improvements in process technology can improve the efficiency of biotechnological industry.

(3)   A wide range of products appears possible both as a result of traditional and new biological methods. Genetic engineering promises the production both of new products, increased yields of existing products and the modification of existing products such as more heat stable enzymes.

Beyond these are less evident factors which will have influenced both investors and economic planners. One is the success of the antibiotic industry. Like other pharmaceuticals this industry has consistently outperformed virtually all other sectors of the economy by whichever parameter is chosen: gross dividends, return on capital, share values and so on. The genetic engineering techniques promise a range of products with similar commercial promise and earning power in interferons, lymphokines and vaccines. Second is the spectacular growth of the data processing industry. Fortunes were made on venture capital investments in the 1960s and early 1970s in embryo computer companies. There is obviously considerable attention paid to the next emergent technology and biotechnology is thought by many to fit the bill. Although somewhat less sanguine than American investors, many national governments are anxious to invest in new technology as a route to future prosperity, most particularly in terms of job creation. The successive oil price rises of 1973/4 and 1979/80 have created recession and unemployment in the OECD countries on a scale not matched since the 1930s. It is widely recognized that capitalist economies tend to pull out of recession by investment in new technology and it is also considered by many western nations that manufacture of the consumer durable products that have dominated their economies since the Second World War is being lost to developing nations. Even Japan, which has been supremely successful in this field, is anxious to develop new industries. There are also strategic considerations at a government level. The OPEC oil crises served to remind many countries of their vulnerability to overseas energy supplies. States were quite simply powerless to act. There is also a widespread belief that oil price rises may be a regular occurrence unless permanent recession or alternative technologies maintain demand at a lower level.

These factors combine to make a sizeable body of reasoning, but it is still insubstantial because of the process limitations which have hindered biotechnological development. Economic analysis is valuable here because it can reveal the cost sensitivity of processes, where improvements need to be made and what impact new methods can have. It can also provide information in areas such as product pricing and availability and price of raw materials.

Political influences

Government intervention has played a critical role in the establishment or retention of much of the fermentation industry world-wide and its influence on the foundation of new biotechnology is crucial. In general the most important influence has been a national balance of payments situation. Governments with a positive balance of payments have tended to retain a free market policy, but when the foreign exchange position is critical they often intervene with actions that can transform the economics of individual industries. Strategic considerations, that is measures to counter a possible adversary, can also be influential.

The actions may be direct, via subsidy or import tariffs, or indirect via the taxation system. India, for example, legislates that its sugar refining industry supplies its fermentation industry with molasses at a fixed price below that of the world market. In the United States fuel ethanol is exempt from some federal and state taxes if it is produced by fermentation indigenously. Brazil charges import tariffs on petroleum to maintain its domestic price above fermentation ethanol. Many of the COMECON countries operate similar protective tariffs to aid their fermentation industries. Government policies of this type are so significant that the fermentation industry would not exist without them in most countries. The political reasoning is generally both strategic and protectionist. South Africa retains an acetone–butanol fermentation plant in part because of its perceived vulnerability to a petroleum embargo. The United States is anxious to free itself from dependence on imported oil from politically sensitive areas of the world as well as finding a use for its agricultural surpluses. Brazil has an enormous foreign debt and must achieve a positive balance of trade to pay it off. Similarly India has a rigorous policy of restricting imports to avoid a balance of payments crisis.

Government policies are not always a positive influence. The EEC has such strong protectionist policies that many critics view the high prices of cereals, and therefore potential fermentation feedstocks within the Community, as being inhibitory to the growth of large scale biotechnology. A European high fructose syrup industry based on imported maize was blocked by effective lobbying by the sugar beet producers. Changes in legislation also put paid to an SCP plant in Italy, though the reasons for this were ostensibly toxicological (Fishlock, 1982).

Most governments have a positive attitude to the development of new biotechnology. There are a number of quite diverse reasons for this in addition to those already mentioned. Both OECD and developing nations are aware that early participation in new technologies is axiomatic for maintaining a leadership and thereby creating future wealth and employment. Historical examples in automobiles, aerospace, chemicals and others reinforce this view. The potential applications of biotechnology are so wide that no-one wants to be left on the starting line. Many nations with agricultural surpluses or the potential to produce them see the possibility of reducing their dependence on imports, particularly oil. Biotechnology also has a good public image, it is seen as making natural products from renewable resources by clean technology. In addition many countries have supported programmes of biological research in universities and research institutes over many years. There have been good reasons for doing so and the outlay has been small when compared to armaments and even particle physics.

Nevertheless there is a strong political desire to cash in on this historical expenditure and form new industries. Hence the emphasis on bridging links and technology transfer from universities to industry. The methods chosen for industrial support depend to a great extent on the political philosophy of the individual nation and its established policies.

National policies towards industrial development

The Eastern bloc countries are known for their centrally managed economies. Japan also plans its economic strategy, but within a capitalist framework under the guidance of the Ministry of Trade and Industry (MITI). This organization has done much to develop Japanese industry along its present, highly successful lines. West Germany relies heavily on capital for new industrial development being provided by industrial banks with different interest rates and loan conditions from the normal commercial banks. All Western European countries tend to have systems of government grants to stimulate the development of new industry. Some, like the UK, offer attractive loan schemes and assistance for new ventures to start up in areas of high unemployment. The US tends to stimulate the development of new industries by a low rate of capital gains tax and favourable tax treatment for R & D limited partnerships. The impact of these different policies will be evident in the subsequent chapters.

The commercial evolution of biotechnology

From an industrial perspective new biotechnology might be an expected outgrowth either of those industries which already employ the techniques (technology led) or of those which make similar products (market led). In the first category some Japanese fermented food companies, notably Ajinomoto, have moved into fermentation, first of amino acids and then of antibiotics, as expansions of their business. In general the European and American craft fermentation industries, for example brewers, distillers and dairies, have not diversified in this direction, although they have been involved in some projects and have supplied capital for new ventures. Gist Brocades, originally a manufacturer of baker's yeast, is a notable exception. They have expanded progressively from this base into supplying products for the baking and brewing industries, then into dairying. They now produce a wide range of enzymes for food, detergents, textiles, paper and the corn starch industries and have diversified into antibiotic production by fermentation. They are the world's second major supplier of enzymes, employing some 7000 personnel (Chapter 7). Their main rival Novo has diversified not from a traditional base, but from a two-man venture to produce insulin in the 1920s. Progressive expansion into trypsin, penicillin, detergent enzymes, amylases and glucose isomerase has made them the world's largest enzyme manufacturer, also involved in antibiotics and with an estimated 25% of the world insulin market. Novo's 1983 earnings exceeded $100 million.

Most pharmaceutical and many chemical companies have entered biotechnology as a means of supplying a product to fit into their range. Antibiotics are the predominant interest of the pharmaceutical companies, but they also manufacture vitamins and some, notably Pfizer, have interests in lower value products such as organic acids. There is a great overlap between pharmaceutical and chemical companies, with many in both businesses, but usually biotechnology is at most a tiny fraction of the interests of bulk chemical manufacturers.

A third category of firms with biotechnological interests may be described as substrate led. Their interest arises from their production of a raw material with biotechnological potential for added value products. Of note here are the American corn wet milling companies (Chapter 9) who make a variety of food products and now fuel ethanol from maize. Some of these companies produce their own enzymes, while others purchase them from suppliers such as Novo and Gist Brocades. Some too are now actively diversifying into enzyme production. In Brazil sugar refiners are diversifying into ethanol production as a means of diverting increased output into a new product which can be more profitable than crystalline sugar.

The final significant group of biotechnological companies are the technology-led new ventures, mostly seen in the US. They have been able to finance expensive research with long lead times by a variety of fund raising methods which are described in Chapter 11.

Summary and outline of the book

Modern biotechnology is a mixture of far-reaching new innovations superimposed on an industry with mixed fortunes. It has many successful products, but has failed to penetrate many market sectors and has lost ground to chemistry in an era of cheap petroleum. The reversals suggest economic disadvantages which are particularly apparent at the low price, high volume end of the product spectrum. The diversity of markets and the novelty of many techniques make economic analysis difficult, but there is a similarity of methodology and available cost and market data which makes some comparisons and conclusions possible.

In this book it is proposed to examine first the influence of the diverse markets, demand and pricing policies on biotechnological products, then the influence of general economic and accounting principles, stressing the importance of capital budgeting and investment in new manufacturing facilities. These principles serve to illustrate the reasoning behind the development of some projects, while others of equal scientific appeal are abandoned. The importance of the production and pricing of raw materials is covered next as it is probably the most significant single input, particularly to the high volume commodities. Fermentation costs are examined in Chapter 5 using data from existing industries and this leads into the problems of downstream recovery and product purification, a critical area covered in Chapter 6. Enzyme catalysis is described in Chapter 7 with special emphasis on the cost saving benefits of immobilization. The important issue of energy and the contributions biotechnology can make to global energy requirements are covered in Chapter 8. In Chapter 9 enzyme catalysis, ethanol production and raw material utilization are tied together in a brief description of the American corn wet milling industry, which may be described as the first example of an integrated process industry very largely dependent on biotechnology. Waste treatment also represents an opportunity for biotechnological processes, some of which are outlined in Chapter 10. The impact of new genetic technology and its extraordinary success in attracting investment in new venture companies are described in Chapter 11. In the final chapter an attempt is made to relate the economic development of biotechnology to the development of other technologies and the current thinking on technology, innovation and economics.

Further reading

Microeconomic Theory. Basic Principles and Extensions, 2nd edn. (1978). W. Nicholson. Hinsdale, Illinois: The Dryden Press.
Economics: Macroeconomic Principles and Issues. (1978). J. Cicarelli. Chicago: Rand McNally College Publishing Co.
Biotechnology – International Trends and Perspectives. (1982). A. T. Bull, G. Holt & M. D. Lilly. Paris: OECD.
Modern Economics. An Introduction for Business and Professional Students, 3rd edn. (1977). London: J. Harvey. Macmillan.
Biochemical Engineering and Biotechnology Handbook. (1983). B. Atkinson & F. Mavituna. London: Macmillan.

2 The markets and industrial organization of biotechnology

Major products and sales

The application of biotechnology spans a range of goods and services, and may be regarded as being opportunistic at present. It supplies certain products either more cheaply or because no other method exists in markets that are dominated by mining, agriculture and chemistry as primary means of production. Whether this base will expand so that biotechnology comes to be regarded as a major means of production remains to be seen. To understand biotechnology in economic terms it is necessary to examine the size of the industry and the nature and relative importance of its markets. The products span a wide spectrum of market conditions, volume, demand and price. The effect of the different markets on production economics is one of the most significant factors shaping the industry.

A 1981 estimate of gross world sales of biotechnological products is given in Table 2.1. In common with all estimates of this type it can only be regarded as an approximation because of currency fluctuations and the factor of so-called soft currencies, which applies to many countries. It is also difficult to value many products manufactured and consumed within one country, for instance Eastern bloc countries, where meaningful exchange rates with the West do not exist and even product tonnages are often unobtainable. It is difficult too to obtain accurate estimates of the world antibiotic sales. Total sales in non-Communist countries in 1975 were put at $4.5 billion (billion being taken as 10^9 throughout) (Klein, 1978). Annual sales were put at $7 billion in 1979 (The *Wall Street Journal*, October 15, 1979, p. 41). Dunnill quotes £4.5 billion for 1981 (Table 2.1). There is also the problem of apportioning the value attributable to biotechnology. Two of the leading groups of antibiotics are the substituted penicillins and cephalosporins. The greater part of their value is in the new side chain which is often synthesized by a series of expensive chemical steps. The biotechnological component is in provision of the bulk starting material and removal of the natural side chain (Chapter 7), and may only be a minor component of the selling price.

Given these reservations it can be seen that biotechnology sales are still dwarfed by the traditional fermentation products, notably alcoholic

Table 2.1. *World markets for biological products, 1981*

Product	Sales (£ millions)
Alcoholic beverages	23 000
Cheese	14 000
Antibiotics	4 500
Penicillins	500
Tetracyclines	500
Cephalosporins	450
Diagnostic tests	2 000
Immunoassay	400
Monoclonal	5
Seeds	1 500
High fructose syrups	800
Amino acids	750
Baker's yeast	540
Steroids	500
Vitamins, all	330
C	200
B_{12}	14
Citric acid	210
Enzymes	200
Vaccines	150
Human serum albumin	125
Insulin	100
Urokinase	50
Human factor VIII protein	40
Human growth hormone	35
Microbial pesticides	12

From P. Dunnill (1984). *SERC* Biotechnology Directorate Newsletter, **1** (1).

beverages and cheese, both of which are still steadily growing in volume. Biotechnology is also still small by the standards of the chemical industry, which has products valued at $700 billion annually (Duncan, 1982) and contributes over 10% of the gross domestic product of most OECD countries (Bull, Holt & Lilly, 1982). The gross sales of biotechnology, excluding alcoholic beverages and cheese, were approximately equivalent in 1981 to the sales of one agricultural product, refined sugar (*c.* $20 billion). Since 1981 there have been changes, notably in fermentation ethanol (not included in Dunnill's estimates) which has doubled to approximately $4 billion in 1984, and in continued growth in high fructose syrups. Antibiotics, ethanol and high fructose syrups now have gross sales in the same range as many of the more important products of the chemical and food industries. Together they dominate sales of biotechnological products (Table 2.2) although there are some clear second rank products such as baker's

Table 2.2. *World sales of major biotechnological products in 1983*

	Volume (1000 te pa)	Price ($ per te)	Value (× $ millions)
Fuel/industrial ethanol[a]			
USA	1000	576	576
Brazil	4500	576	2600
India	400	576	230
Others	400	576	230
Total			3636
High fructose syrups			
USA	3150	400	1260
Japan	600	400	240
Europe and elsewhere	200	400	80
Total			1580
Antibiotics[b]			
Penicillin & synthetic penicillins	—	—	3000
Cephalosporins	—	—	2000
Tetracyclines	—	—	1500
Others	—	—	1500–2000
Total			8000–8500
Other products			
Citric acid	300	1600	480
Monosodium glutamate	220	2500	550
Yeast biomass	450	1000	450
Enzymes	—	—	400
Lysine	40	4000	160
			2040
Total			15256–15756

[a] Based on the US price ($1.70 per gallon).
[b] Approximate estimates from the data of Klein (1978) and *The Wall Street Journal* (15 October 1979, p. 41) corrected for inflation.

yeast, citric acid and monosodium glutamate, with lysine sales rapidly rising into this category. It is also important to bear in mind the rapid growth of fuel ethanol and high fructose syrups (Chapters 8 and 9). If these tables had been drawn up as recently as the mid-1970s neither product would have had a significant showing.

The products in Tables 2.1 and 2.2 fall clearly into categories of medicine, food and industrial ethanol, with a very minor contribution from microbial pesticides. Even the products of recombinant DNA technology (Chapter 11) are still predominantly aimed at these markets. Ethanol stands alone as the only bulk commodity used as a fuel and chemical feedstock. It is also the only product which is competing directly with an identical product of the petrochemical industry and is

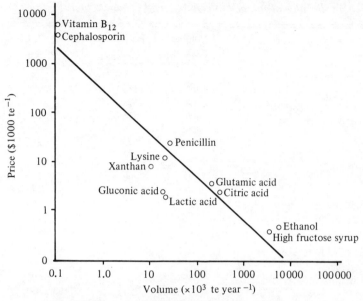

Fig. 2.1. Hierarchy of values for biotechnological products.

marketed in the US at a competitive price, although subsidies are involved. High fructose syrups similarly compete with a virtually identical product, inverted sucrose syrups, obtained by agricultural production and chemical hydrolysis, but again there is government intervention (Chapter 9). The other products are all either too complex for synthetic routes or they are food ingredients in which substitution by synthetic products is not permitted. This provides the first evidence on the relative costs of biotechnological processes. They tend only to compete in high value markets in medicine or foods where a premium is paid for natural products with an absence of toxic impurities or suitable taste characteristics. In terms of price, only ethanol, high fructose syrups and baker's yeast lie below $2000 te^{-1} in the range occupied by bulk commodity chemicals such as alcohols, aldehydes, acids, feedstocks for polymer synthesis, bulk surfactants and bulk food or feed additives such as sucrose, starches, soy bean and fish meal. Single cell protein has not been included in this category because its production is not viable under the prevailing circumstances in Western economies.

The relationship between price and total sales volume can be represented by a double log plot (Fig. 2.1) when many of the products lie on a straight line. Similar plots can be drawn for other products, for example thermoplastics (King, 1982). Several products lie below the line, i.e. their sales volumes are lower than would be predicted from the price, indicating a relatively low utility. Conversely ethanol lies above the line, which is probably more a consequence of tax credits making it viable for bulk use as a fuel than greater utility. Without the tax factor in the US and Brazil, ethanol at approximately $1.70 per US gallon (or $560 te^{-1}) would not be incorporated in petroleum blends.

Precise interpretations of plots of this kind are suspect because of the scatter of points and subsidy factors, but they do emphasize the necessity of reducing production costs to achieve bulk sales. The progress of the products of the chemical industry down the price scale and their concomitant penetration of bulk markets are good illustrations of this point (King, 1982). Amino acids also provide an example of this phenomenon. Most amino acids are used only in infusions and as therapeutic agents. They have world sales below 200 te per year (Hirose & Okada, 1979) and are expensive. Where other uses such as food or animal feed additives have been found, annual production has risen and prices have fallen, but only where the technology has permitted. This is illustrated by monosodium glutamate (Table 5.1), lysine (Tosaka, Enei & Hirose, 1983) and methionine (synthesized chemically) which had sales of 450000 te in 1982 at $2500 te^{-1}. Tryptophan, another essential amino acid, could also find bulk applications in animal feed, but only if prices drop from the current $100 kg^{-1} to $10 or less. Genetically engineered strains may make this possible. Similarly there is now pressure to reduce the costs of phenylalanine now that it has larger sales potential in the manufacture of Aspartame L-aspartic acid-L-phenylalanine methyl ester (a new high intensity sweetener).

Categories of products in terms of volume and value

The prices of biotechnological products range from glucose and high fructose syrups at $300–400 te^{-1} up to vitamin B_{12} or some antibiotics at $8 million te^{-1} or higher. It is convenient to split these products into three groups (Table 2.3 and below) although the classification is arbitrary and in some respects the borders are blurred. Penicillin, for example, is regarded by the pharmaceutical industry as a low priced product, whereas in relation to the organic chemical industry it is expensive. It can also be seen in Fig. 2.1 that the products fall into these three groups and that penicillin is in the middle category (in bulk unsterile form). As the range of products obtained by biotechnology increases, the borders between the groups will inevitably become blurred and many important areas such as microbial pesticides, enhanced oil recovery or mineral leaching already cannot conveniently be incorporated into this scheme. Nevertheless this categorization provides a convenient starting point for the analysis of production costs and markets.

Table 2.3. *Biotechnology: based on volume and value*

Category	Activities
High volume, low value	Methane, ethanol, biomass, animal feed, water purification, effluent and waste treatment
High volume, intermediate value	Amino and organic acids, food products, baker's yeast, polymers
Low volume, high value	Antibiotics and other health care products, enzymes, vitamins

From Atkinson & Mavituna (1983).

The first group comprises high volume, low value bulk chemicals or foods such as ethanol, acetone, butanol, biomass (SCP) and high fructose syrups. It is in this group that competition either with petrochemicals or agriculture is most intense and where most ground has been lost to the chemical industry. It is also where most money has been lost in biotechnological projects, notably in SCP. Even with the high oil prices after 1980 most of these biotechnological processes are not feasible without subsidy. It is in this area that the broadest aspirations of biotechnological development, i.e., chemical feedstocks and fuels from renewable resources, must be met.

The second group comprises lower volume, but still relatively cheap commodities, generally referred to as speciality chemicals, including many organic acids, the amino acids, polysaccharides and possibly other polymers in the future. Here too in some cases inroads have been made by the petrochemical industry, and the biological product may have survived only because of its use in foods where either a 'natural' product only is permitted or will attract a premium price, or it is too expensive to purify the chemical product sufficiently. For example, most food grade lactic acid is produced by fermentation and commands a higher price than the industrial grade product derived from lactonitrile.

The third group of low volume, high value products is composed of goods that can only be obtained by biological routes: antibiotics, vitamins, enzymes and vaccines. Even so some products have faced competition from chemical synthesis, most notably riboflavin, where most non-animal feed product is made chemically, although the precursor, ribose, is still made biologically. This group represents to many the most profitable area for expansion of biotechnology, at least in the short and medium term, and it is here that most of the initial targets of recombinant DNA technology have been identified.

Demand and price

In economic terms the relationship of prices and uses may be defined more appropriately in terms of demand curves. The categories of biotechnological products may then be described in terms of elasticity of demand. General economic equilibrium theory states that demand for commodities and all resources depends on prices. Supplies of commodities from firms depend on commodity prices, the coefficients of production and the prices of resources. The demand for each commodity equals its supply, all demand and supplies being compatible because of interconnections through the network of prices (Watson & Holman, 1977). Although this theory has been criticized as being a complex statement of the obvious it is a useful approach to explaining product pricing and it is relevant to understanding the successes and failures of different biotechnological products.

Demand curves are plots of price *versus* quantity sold. The slope of the plot (generally negative) is the price elasticity of demand. Different types of demand curve are shown in Fig. 2.2. If the plot is vertical demand is perfectly inelastic, i.e. demand is completely insensitive to price. If it is horizontal it is perfectly elastic, but in fact these two extremes are never reached. Unit elasticity is the mid-point where a 1% reduction in unit price produces a 1% increase in demand rate. A slope of less than 1.0 is described as showing an inelastic demand and a slope greater than 1.0 is described as an elastic relationship. Highly elastic refers to demand highly responsive to changes in price. Negative elasticity, i.e. demand decreasing with a fall in price, is confined to a few exceptional products (such as diamonds), none of which appears to be related to biotechnology, although some products sold in health food stores may fit the bill.

In general the bulk products of biotechnology all show high elasticity of demand (Fig. 2.2a) because they are challenged by other products or other methods of production. Single cell protein is perhaps the example with the highest elasticity because it is competing with other products, notably soy and fish meal which can substitute more or less directly. The intermediate group of products, organic acids, amino acids etc., while not having exactly unit elasticity, approximate to this position. They do not have direct substitutes, but demand in food or feed is still price sensitive. Most pharmaceuticals, including antibiotics, are good examples of highly inelastic demand, particularly when the drug is new and has no effective substitutes. In health care the patient's welfare is considered the first priority, largely irrespective of price. The three categories of biotechnological products may therefore be redefined in terms of high, medium and low elasticity of demand.

In reality the situation is not as clear cut as this suggests. The curves

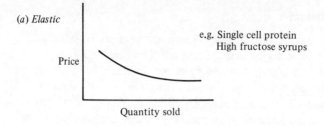

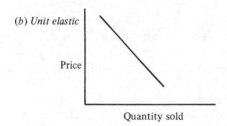

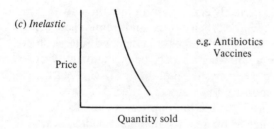

Fig. 2.2. Demand curves.

for many products have changing slopes at different points and they may have a sharp change, i.e. be biphasic, at a critical point where the price of a direct competitor is equalled. For example the demand for antibiotics in human medicines may be inelastic, but inclusion in animal feeds (where permitted) and in some industrial processes may be highly elastic. As a potable product fermentation ethanol is generally not challenged by the chemical product, and its demand curve is fairly inelastic, but below a threshold price it becomes competitive with synthetic ethanol for industrial users. At this point there will be a sudden increase in quantity sold and demand will become much more elastic. This situation is represented in Fig. 2.3.

Research expenditure has been related to elasticity of demand. The higher the elasticity the lower the risks which are deemed acceptable and the lower the research expenditure. The high inelasticity of demand of pharmaceuticals is one of the major factors which supports high

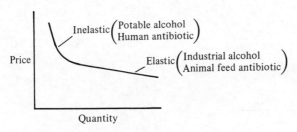

Fig. 2.3. Biphasic demand curve.

levels of research. The expenditure can be recouped in the price if the product is successful. A particular problem which besets an innovative industry is the estimation of price that a user is prepared to pay for a product of which as yet he has no experience. The industry may have to convince the customer to want and use the new product. It must be stressed that evaluation of chemicals to an industrial user is entirely objective. The intrinsic value of the work, raw materials or skill is of no concern to the customer as it might be in consumer markets such as furniture or cars (Bradbury & Dutton, 1972).

The general principles underlying elasticity of demand in biotechnology (as elsewhere) may be summarized as follows:

(1)   The number and closeness of substitutes.
(2)   The importance of the commodity to a buyer.
(3)   The number of uses. If there are only a few uses demand is often inelastic. The more uses, the more elastic is the demand. The uses usually stand in a hierarchy, as the price is lowered more uses become possible. For example, a hierarchy for ethanol would be alcoholic beverages > industrial solvents > cleaning fluids > fuel.

Price is not the sole consideration of a customer. Other factors such as quality and reliability of supply are important. Customers will often not switch suppliers for a marginal price cut; longer term questions of supply and future pricing are always considered. In particular, major industrial customers will maintain more than one supplier for reasons of reliability and potential vulnerability, even if this means paying more in the short term.

Supply

In the same way that demand curves may be drawn relating quantity of goods bought to price, supply curves showing quantity of goods which are produced and sold at any given price can also be constructed. These curves have a positive gradient, the higher the price the greater the production. There is a threshold price below which it is not worthwhile to supply, since minimum production costs cannot be met (Fig. 2.4*a*).

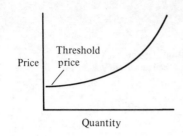

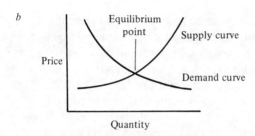

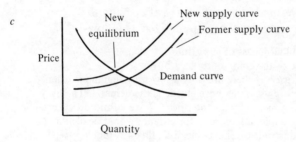

Fig. 2.4. Supply curves.

The supply and demand curves meet at an equilibrium point which is altered if the price is increased or decreased. A family of such curves can then be constructed at different price levels. The laws of supply and demand are then reduced to this equilibrium concept. The equilibrium price of a good is the price at which supply and demand are equal (Fig. 2.4*b*, *c*).

Again this is a simplistic treatment, but it is useful in explaining many forms of economic behaviour. One limitation is the time taken to build up production, whether this involves construction of new plant or planting additional crops in agriculture. If production is built up it often leads to economies of scale, hence the price may fall, thus increasing demand and leading to short, medium and long term equilibria.

Price elasticity and supply curves are major considerations in capital investment decisions which must allow for repayment of investment over a number of years. For example the large investment in ethanol fermentation facilities by US corn millers has only taken place with the guarantee of fuel tax rebates. These ensure that fermentation ethanol use in petroleum blends (Chapter 8) is attractive for a given period of time. Otherwise it would be too risky, even though production costs of ethanol by corn wet milling in particular match the present price of synthetic ethanol. If the price were to drop, either as a result of oversupply or competition from synthetic producers, the fermentation plants would become non-viable. The equilibrium prices would fall below the threshold price for the fermentation route.

Biotechnology may suffer, as has the chemical industry, from periodic price drops due to overproduction. Manufacturers anticipating rising demand build big new plants to keep marginal costs down. This can then lead to oversupply, particularly when coupled with an economic recession. Corn wet millers have suffered from this problem, but have countered it by switching to new products (Chapter 9).

Industry sectors

General supply and demand theory goes some way to explaining prices, sales volumes and the successes and failures of some biotechnological products, but it is insufficient for a detailed understanding of the markets. Another factor that must be understood is the characteristic of each market area or industry involved. For example, as described in Chapter 5 the economic factors in the production and marketing of industrial and potable ethanol are sufficiently dissimilar for them to be regarded as different businesses despite biochemical and process similarities. The most important market sectors being supplied by biotechnolgy are pharmaceuticals, food, agriculture, chemicals and fuels. Each has very different characteristics which play an important role in dictating strategies in production, marketing, pricing, research and development.

Pharmaceuticals

The total world-wide sales of pharmaceuticals has been estimated at $80 billion in 1983 of which approximately 10% is produced, in part at least, by biotechnology. This is predominantly antibiotics, but also includes contributions from vaccines, steroids, vitamins, amino acids and enzymes. This contribution is likely to rise rapidly when the products of recombinant DNA technology come on stream. Pharmaceuticals are nearly all at the upper end of the price range, but prices on a weight basis are misleading in view of efficacy. A cephalosporin, retailing at $6 million te^{-1} works out at $4.50 for an average 750 mg injection. A patient suffering from even a mild infection is unlikely to complain (on a price basis at least) if he requires four injections costing $18 in total. It is true that the cost of medicines on a mass basis, particularly vaccines, to developing nations cannot be dismissed quite so glibly, but nevertheless on an efficacy basis many drugs, especially the older established ones, cannot be regarded as excessively expensive.

The cost structure of pharmaceuticals is quite unlike other products manufactured by similar technologies. As with potable ethanol, manufacturing costs generally represent a relatively low proportion of the selling price, averaging under 50%, but for different reasons. The nominal cost of raw materials in pharmaceuticals manufacture is often under 10% of the selling price. The position of the consumer in ethical pharmaceuticals is unique. The doctor decides what he considers to be in the patient's best interest and selects the best drug therapeutically, irrespective of cost. This leads to the widely held view that demand for pharmaceuticals is highly inelastic, which is true for new drugs, but less so for older ones. A drug's sales are determined by the extent to which the manufacturer succeeds in gaining 'doctor acceptance' of his product (Cooper, 1966). This is achieved by intense and expensive marketing, often through technical sales literature, conferences, symposia and salesmen canvassing individual physicians. Total selling costs for pharmaceuticals have been estimated to be in the 30–45% of selling price range (Walker, 1971; Egan, Higinbotham & Watson, 1982). In fairness this does include distribution of essential information to the medical profession. Rivalry in the pharmaceutical industry has therefore been diverted to some extent from price competition to product differentiation. This has stimulated a high rate of innovation and high expenditures on research and development. It is estimated that overall the industry spends 12% of total sales revenue on research (Rapoport, 1983). Research and development has become a primary vehicle of competition in the industry (Cooper, 1966; Walker, 1971).

This does not mean that there has not been pressure to cut manufacturing costs. In fact as described for penicillin, there have been spectacular improvements in productivity and reductions in price. The original cost of $200 g^{-1} has dropped to 3–5 ¢ g^{-1} today. The degree of price competition varies considerably with the class of drug and the individual product. A new product offering unique therapeutic advantages and protected by patents has a relatively inelastic demand curve and can earn a price premium. (Penicillin is atypical in that it was never covered by a patent.) Dependent upon the nearness and efficacy of the competition, manufacturers are free to fix their prices during the period of patent protection. As time passes and substitute products are found by competitors and patent expiry approaches, the manufacturer must devote ever increasing efforts to cost reduction and maintaining his market share once patent cover is lost. While the industry does compete in price, and there are often relatively high cross-elasticities of demand between drugs in the same group, there is also a high element of non-price competition. Besides innovation this takes on many aspects of quality, such as efficacy, safety, clinical evidence and experience, information to physicians and the reputation of the manufacturer. The high rate of innovation has a corollary in that the life of many products is short. In these cases only limited returns are to be gained from process improvements or economies of scale. In general this has not been true of antibiotics, which have had long lives, and it may not be true of newer biotechnological products. Perhaps as a consequence of many products having a short life, however, most pharmaceutical plant is designed to be adaptable, and flexibility is often obtained at the expense of economic efficiency in the production of a particular product.

Pharmaceutical research has traditionally been based on empirical compound testing and clinical observation. Increasingly feedback from new basic knowledge of biological processes is being applied to the search for new drugs. This is especially true in the field of recombinant DNA. All new drugs are subject to rigorous regulatory control. Manufacturers must file applications for approval at an early stage of development and regulatory authorities (particularly the Federal Drugs Administration in the US) are involved in clinical testing and must approve the design of experiments. This process is slow and expensive. The average clinical development time for a new pharmaceutical is in the order of 6–7 years and the preclinical stages before this will have taken several years. The cost of development of a new pharmaceutical is generally put at at least \$100–150 million. These costs must be recouped along with other costs during the years of patent protection. The discovery is far less costly and time consuming than the development of a marketable product, but of course a large number of unsuccessful products must be screened for every successful one. Companies must have a critical mass of research and development organization to ensure one or two big successes among many failures. Some even very successful companies may only have a limited number of big selling products. As a result there are high entry barriers to new firms in the pharmaceutical industry; instead it tends to be dominated by a small number of large firms world-wide.

The pharmaceutical companies have been very profitable. In the US in the period 1958–75 the rate of return for all manufacturing averaged 11.0% compared to 18.1% for the drug industry as a whole and 19.7% for the 12 largest companies (Measday, 1977). In Britain too pharmaceutical companies have been stock market leaders and the pharmaceutical divisions of chemical companies have outperformed the other sectors. Few if any other industries in the economy can match their record, and the high profit rate has become the most controversial characteristic of the drug industry. The high profit rates have however been associated with high risk products and product innovations have been the cornerstone of the industry's success. In welfare terms the achievements of the drug industry in reducing mortality and suffering by developing new products have been considerable.

The food industry

If the traditional fermentation industries are included the food industry dominates biotechnology. Even without them it is still the largest area for biotechnological processes after pharmaceuticals in value terms, and probably the largest in volume, although this judgement is dependent on the relative sales of high fructose syrups and fuel ethanol. Considerations of taste have often favoured the retention of traditional processes despite production economies that could be made by the adoption of new techniques. The promise of biotechnology is that process improvements can be made using techniques which permit the retention of desirable organoleptic properties and are considered 'natural'.

The food industry is huge in terms of turnover and labour employed and is very diverse, ranging from small individual producers to giant multinational corporations. It also occupies an important position in economic activity. It receives inputs from the chemical, paper and pulp industries as well as the principal one from agriculture and can exert a pull effect on these other industries through its demand for raw materials. There are of course strong links with agriculture and a great deal of overlap.

In developing nations food accounts for a high proportion of total employment and total manufacturing output. It accounts for upwards of 40% of all household consumption and mostly involves small individual producers. In industrial nations food expenditure still accounts for 20–30% of household consumption and there is increasing dependence on the food processing industry to package, preserve, transport and distribute food to centres of population. There is a tendency towards increasing scale with large processing plant and large manufacturing companies. Such food plants are highly automated and capital intensive. The food industry in Europe and North America has large, vertically integrated firms whose operations extend from agriculture through the processing factories to supplying packaged products to the retail industry. It is in these big food processing firms that biotechnology is having most impact. These firms have the technology and scale to incorporate new biotechnological processes, particularly based on enzymes. The starch industry in North America, for example, is a large scale user of biotechnological processes.

In contrast to pharmaceutical companies most large food manufacturers are high volume, low margin operations and their profits based on gross sales are often very low. Research and development expenditure is a very much smaller proportion of turnover. Activities have tended to concentrate on minor changes in product quality and attributes rather than on basic technology. Market research has become more significant than basic research. However, many large food processing firms are multinational and able to afford expensive research programmes because they can spread the benefits world-wide. If the cornstarch industry is considered as an example much basic research is more likely to be performed by the biotechnological industry suppliers, for example enzyme manufacturers, rather than by the food companies themselves.

Demand for food products is elastic, particularly at the bulk end of the market. As will be shown in Chapter 7, processes such as high fructose syrup manufacture are highly cost-conscious and are finely tuned. Many products supplied to the food industry lie in the middle price group, for example organic acids, amino acids and xanthan. Very few high price products are likely to be viable. They would be confined to flavour and aroma compounds; saffron, for example, has been suggested as a potential product of plant tissue culture. Even the new high intensity sweetener Aspartame is predicted to fall in price from *c.* $80–90 per lb (1983) to *c.* $20 per lb or less in the next few years. Enzymes for the food industry do come in the upper price category, but they must be rugged and re-usable, usually by immobilization.

New food products must undergo a similar programme of regulatory approval to that demanded for pharmaceuticals. Approval of Aspartame has taken over ten years. The research and approval of Rank Hovis McDougall's *Fusarium* SCP food Mycoprotein was estimated to have cost the company £30 million (Fishlock, 1982). The use of some enzymes in food processes is prohibited because the producing organism does not have regulatory approval and it is not considered worthwhile to acquire it. Food is also highly political. High fructose syrups have only achieved their present sales in the US and Japan because of tariffs or taxes (Chapter 9). Their spread to Europe was blocked by levies.

Wastes from the food industry, especially from large processing plants, are becoming an increasing financial problem to food companies because of stricter legislation on the dumping of high biological oxygen demand (BOD) wastes. This is presenting a useful market for biotechnological waste treatment systems, many of which can now generate valuable by-products as well as achieving waste removal.

Agriculture

In many countries agriculture is still dominated by small individual producers using traditional techniques in a free competitive market. In Western countries however, agriculture is very advanced in terms of crop breeding, strain selection, fertilizers, pesticides, animal husbandry and mechanization. It covers small farmers and large agribusiness combines, but is increasingly capital intensive with decreasing labour inputs. The impact of biotechnology is predicted to be chiefly in the agriculture supply industry, for example in new plant strains, fertilizers and pesticides. There are also considerable market opportunities in animal husbandry. Antibiotics are already used widely in animal feeding. Near term prospects include SCP feeds, vaccines against diseases such as swine fever or foot and mouth, and growth hormone. In the longer term genetic improvement of strains may be developed.

Agriculture is another politically sensitive area. In addition to trade barriers and protectionist policies, governments frequently underwrite research programmes and are likely to provide extensive funding in subjects such as new crop varieties. There are many popular misconceptions about agriculture. It is a very cost conscious, innovative and often efficient industry in many countries and has seen enormous increases in productivity in recent years. It provides many prime market opportunities for biotechnological products, but is a hard area to sell into, in that products must have a high degree of demonstrable efficacy.

Chemicals and fuels

The chemical industry is huge and diverse. It includes bulk products such as fertilizers, polymers, solvents, dyestuffs and surfactants through to consumer products such as perfumes, polishes and paints. Its products span all three price/volume categories. The industry grew from the 19th century on coal and to a lesser extent on wood as feedstocks. In America the move to oil took place in the 1930s. Europe followed later, after the Second World War, but since then petroleum has been the universal raw material. In Britain for example, 11% of all organic chemical production was based on oil in 1949, while by 1968 it was 87%. Originally polythene (polyethylene) was made on a small scale with ethylene costing £250 per ton derived from fermentation alcohol. The breakthrough in production was the switching to petroleum-based ethylene at £90 per ton in 1952 (Reuben & Burstall, 1973).

Throughout the 1950s and 1960s petroleum remained cheap, declining in actual as well as real terms. Coal by comparison rose in price because of its high labour content (35% of total costs in the US

and over 50% in Europe). In addition to feedstock price, improvements
in process efficiency and economies of scale caused the price of all
petrochemicals to fall steadily until 1973. The growth of the chemical
industry in this period was explosive, in some years approaching 20%
per annum in certain industrial countries. It is an efficient, capital
intensive, highly automated industry with high wage levels. Labour
costs only represent about 10% of total costs. The scale of operations
can be judged from the fact that ethylene crackers are now in the order
of 300–500000 te per year. About 80% of the output of the world
organic chemical industry is used in the production of synthetic
polymers. There has been a tendency towards vertical integration: a
form of business organization in which one firm generates successive
processes of manufacture or distribution of a product. It often includes
raw material production and distribution, for example oil,
petrochemicals, plastics, paints etc. Many oil companies have diversified
into chemicals. There has also tended to be a concentration in OECD
countries (Japan, North America and Europe), but this may well move
to the Middle East where new plants are being constructed in oil
producing areas.

Since the oil price rises there has been a fall in demand. The industry
now has substantial overcapacity in some areas and has suffered quite
badly in recent economic recessions. The emergence of new large
tonnage materials has dropped off in recent years. It is a mature
industry that can be regarded as standing at the end of a period of
growth and rapid technical development which has been dominated by
petrochemicals and polymers.

The prospects for biotechnology lie in replacing the chemical industry
as a means of production using biological feedstocks, in supplying
feedstocks such as ethanol from renewable resources for chemical
synthesis and in supplying specific compounds which can be made more
cheaply by biological means. At present the contribution of biological
processes to the chemical industry (excluding pharmaceuticals) is very
marginal: some steroid modifications are carried out enzymically,
itaconic acid is supplied on a limited basis for the manufacture of some
thermoplastics, ethanol is used as a feedstock in India and Brazil, and
acetone and butanol are manufactured by fermentation in South Africa.
There have been many bold statements on renewable resources, but the
task is daunting when the scale, efficiency and low costs of much of the
present chemical industry are taken into account. Biotechnological
processes have been displaced by chemistry in some cases because they
are less efficient and more expensive, but there are grounds for
optimism. Biological feedstocks are falling in price relative to oil,
fermentation ethanol has become relatively less expensive, new genetic
techniques can supply new products and new reactor designs can offer
process improvements. In other areas, for example fertilizers,

biotechnology may displace the use of chemically synthesized products indirectly. There is however a long way to go, and in the mean time biotechnology will probably benefit more from the lessons of the chemical industry on developing new products and finding markets for them.

The development of biological fuels faces a similar set of circumstances to chemical feedstocks, but there are some distinctions. First it can be argued that oil is more valuable as a feedstock than as a fuel and alternative fuel development should be given more priority to conserve oil as a feedstock. Solid fuels, notably coal and natural gas, can replace oil in many heating applications and in electricity generation. Nuclear energy can be used for electricity generation. There is then a need for liquid fuel (hence the interest in ethanol), but there are competing alternatives such as methanol and coal liquefaction. Secondly, fuel has an important political dimension. Governments effectively regulate supply through taxation, which in most countries is 50% or more of the selling price. Thus although fermentation ethanol cannot compete directly on price at present, the use of selective taxation policies, direct subsidies or import controls can stimulate its production. As in agriculture or food, direct cost factors may be overridden by political considerations.

Oil is dominated world-wide by multinational conglomerates which are highly efficient, high technology organizations committing vast resources to new exploration and exploitation of oil. They have considerable interest in biotechnology. Many were originally involved in SCP production from petroleum, while many are now suppliers of capital for new biotechnology ventures in addition to extensive in-house research. Their interest in biotechnology also includes enhanced oil recovery and extraction of oil from shale and tar sands.

Other markets

Most applications of biotechnology are covered by the categories listed so far, but there are others which are difficult to classify and to analyse economically. For example mineral leaching of ores by microbial oxidation is already used by some mining companies for the recovery of more valuable metals such as copper. Waste disposal or recovery covers a variety of industries plus public health. Its benefits are often in public welfare and are difficult to quantify. Biological equivalents to microchips and semiconductor technology are at present a highly speculative area, whose impact cannot be guessed at. On a more practical level *Pseudomonas syringae* strains which cause frost damage to certain crops are claimed to function as ice-nucleation points in artificial snow-making machines.

Industry structure and forms of competition

Analysis of the biotechnology industry so far has covered products and markets. The third important characteristic is the organization of producers. This is generally classified according to the number of sellers in the market. Accordingly industries are classified as competitive, oligopolistic or monopolistic. The fewer the number of sellers the greater the market power conferred. Once again the very diversity of biotechnology, its status as a means of production and the markets supplied will ensure that all categories are covered. It is also difficult to distinguish whether biotechnology or the market area, for example pharmaceuticals or chemicals, will be the stronger determinant in this organization. Nevertheless it is interesting to look at the structures that have already emerged in various sectors of the industry.

The biggest single determinant of the number of producers is usually the barriers to entry of new firms. These include brand loyalty to existing firms, penetration costs such as advertising to overcome lack of product recognition and promotional expenditures. In biotechnology as elsewhere the main barriers are research and development costs, the maintenance of a high level of technological expertise and capital outlay on new plants with sufficient economies of scale and process efficiency to compete in the market. Once a fermentation industry for a particular product is established and process improvements have been made (Chapter 5), it is very difficult for a new entrant to match these and compete. There is a great premium for being the first or a very early entrant in a field. Later firms find it difficult, if not impossible, unless the market is expanding very rapidly or they acquire an already established producer. This also extends to fermentation technology in general. Even if the new firm has a new product, it is difficult and costly to acquire the technical expertise in factors such as construction of plant, maintenance of asepsis, downstream recovery and so on. Other barriers to entry include patent costs and particularly regulatory costs in new drugs and foodstuffs. In these cases a new small producer is under much pressure to form a partnership, even if on a limited basis, with a large company in order to gain regulatory approval (Chapter 11). Large companies also tend to form partnerships in these areas to spread costs and risks of new products.

Competition between established producers is different according to the market sector involved. Pharmaceuticals occupy a unique position because of their supply by prescription. Also in some countries with state controlled medical services a single price of supply to the health service is negotiated, although the health service does not influence doctors' decisions to supply the drug and therefore does not control the volume of sales. With many products competition is not just between biotechnological manufacturers, but between biotechnology and chemical routes to synthesis, as in lactic or acetic acids, acetone and butanol, or agriculture as in SCP. Competition in these cases is not just a matter of efficient production, distribution and marketing, but is dependent on the relative costs of the two routes and is frequently determined by costs of raw materials or political intervention.

Pure and perfect competition

The presence of many small firms trading openly and competitively in a free market situation, as say in agricultural produce markets, is relatively rare in biotechnology because of the entry barriers. There is however strong competition in the supply of diagnostics, clinical kits, monoclonal antibodies and some small sales volume enzymes where a large number of new firms are supplying products and competition is based largely on product differentiation and innovation rather than on price. It is a rapidly evolving sector of biotechnology and may come to be dominated by large companies in the future, but at present capital costs are low and there are relatively low barriers or entry to innovative and highly skilled individuals or groups with new products. (This is not intended to minimize the amount of perseverance, energy and determination required to set up a venture and enter the market.)

Monopoly

A monopolist is defined as the only manufacturer of a product which has no close substitutes. The emotive connotations to its use are ameliorated somewhat by the fact that most monopolies in biotechnology are conferred by patents. They are therefore only temporary and justified to recoup development costs and stimulate innovation. Monopolies outside patents are fairly common, but are often brought about by a series of factors. For example Merck are now the sole manufacturers of riboflavin by fermentation in America, although there were formerly five or six producers. The reason is competition from the chemical synthesis route via ribose. The only economic outlet for the fermentation product is in animal feed and this market only supports one supplier. Likewise there was only one major producer of xanthan until recently. They were first in the field with the product and the high cost and technical obstacles in developing a process and building a plant to enter this relatively small market had deterred successful new entrants. The steadily increasing sales of xanthan have since attracted new producers. There are other relatively low volume biotechnological products such as ribose which also have only one producer.

Monopolists maintain their position because of market size, their

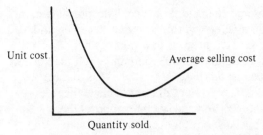

Fig. 2.5. Effect of selling expenses on unit costs.

technological lead (initially protected by patents) and the investment required by a new manufacturer. New entrants must always beware of the established producer cutting prices below an economic level to prevent them gaining a foothold (predatory pricing). Monopolists work to maximize profit, not price, and are under pressure to improve efficiency. This in turn may give them an unassailable lead. In areas such as biotechnology where innovation and research are so important monopolies can have some beneficial effects. Indeed the view has been expressed that monopoly power is necessary to justify substantial expenditure on research and development (Schumpeter, 1942). In general, although monopolistic prices are higher than competitive prices, they are not that much higher. In time monopolistic power is usually lost either by technological changes, patent expiry or by market growth being sufficient to attract new entrants.

A form of monopoly, monopolistic competition, also exists in pharmaceuticals. Here each firm has a differentiated product with stated advantages, although overall the products compete on many points. This is common in antibiotics where many different types of cephalosporins or semi-synthetic penicillins are marketed by different firms. Much of the competition is then of the non-price variety, depending heavily on advertising and brand promotion. A different type of cost *versus* volume curve applies in this type of situation (Fig. 2.5). The initial decline is due to economies of scale, but the mounting cost of selling expenses takes over and the selling price may actually rise as a consequence.

Oligopoly

The majority of large sales volume biotechnology products such as amino acids, organic acids, bulk enzymes, older antibiotics, and high fructose syrups are manufactured by oligopolies. These may arise by new firms entering the market with a product once the patent cover of the initial manufacturer has expired or through the loss of small producers because free market competition proved too intense. For example, in the case of the older antibiotics, such as streptomycin or some tetracyclines, once the initial patent expired several other antibiotic manufacturers entered the field. Some cephalosporins present a different case where the discoverer, via the National Research and Development Corporation, granted licences to a number of manufacturers world-wide. The glut of small companies manufacturing the microbial insecticide *Bacillus thuringiensis* in the 1960s and early 1970s proved to be too great. The product did not achieve the sales expectations of the manufacturers and today only supports three companies. Another oligopolistic situation occurs in American corn wet-milling (Chapter 9). Similar situations are found in most high technology industries with high capital investment and high entry barriers, such as oil, computer manufacture and automobiles.

The economic theory of oligopoly is more complex than pure competition or monopoly and is best covered in specialist texts. In general, oligopolies exhibit a type of group behaviour; there is competition, but there are often rules or codes of conduct. Firms often have interdependent cost curves because they compete for resources. They also tend to guess the actions of their rivals in response to their own, which leads to strategies and game playing. Oligopolistic prices may remain unchanged for some time, unlike in free competition situations. Often one firm sets a trend and others follow. The most extreme form of oligopolistic behaviour is found in the formation of cartels, where an explicit agreement exists between independent producers on price, output and sometimes on division of sales and territories. A perfect cartel results in the maximization of the joint profits of producers, but cartels are often short-lived and break up in mutual distrust or a squabble over distribution of profits. Cartel formation on the fixed lines of say IATA in the airline industry or OPEC in oil has not occurred in biotechnology. In other respects many forms of oligopolistic competition occur.

As in the monopoly situation pricing structures and large capital requirements provide strong barriers to the entry of new firms into the market. They will also have high overheads due to their new plant, but there are exceptions when an existing firm can use surplus capacity, as may happen in fermentation where equipment is fairly flexible. A new firm may also enter a market when the existing producer(s) have reached full capacity and demand is increasing.

Determination of price

In general prices are set to maximize profit, for example in unit costs *versus* number of units considerations. The dependence of prices of biotechnological products on costs – raw materials, labour, plant construction etc. – will be dealt with in future chapters. The effects of demand and supply have already been described. Other factors also have a bearing. The price must reflect the re-investment of firms in new research and development as well as plant construction and cash flow in new ventures. It is also often set to dissuade new entrants into a market and may even be lowered below cost to prevent this if not prevented by legislation. Research scientists often believe that they can undercut existing prices until such factors are taken into consideration.

Another factor is price discrimination. A firm may charge two or more prices for the same product at the same time, depending on quantity, location of buyers, use of product, evenness of demand, future contracts and other factors. This is due to the existence of separate markets with differences in elasticity of demand. For example a domestic market may be protected by tariffs, but a lower price may be accepted in an export market. In Britain pharmaceutical companies may obtain higher prices for drugs than in many export markets (Erlichman, 1984). Penicillin has different prices for industrial, animal feed or human drug use. Different grades with different purification costs also influence these prices. Food grade products also often fetch a higher price than is accounted for by production costs alone.

Summary

The wide range of products made by present day biotechnology may be arbitrarily put into three classes based on volume and value, though it is more accurate and informative to classify them in terms of elasticity of demand. The differences in demand place quite different constraints upon the economics of production. Biotechnology products are principally concentrated in four market sectors: pharmaceuticals, food, agriculture and chemicals/fuels. Again these areas impose different conditions on producers.

In addition to the range of market conditions, different biotechnology sectors cover all forms of competition and industry structure from pure and perfect competition through oligopoly to monopoly. The older business areas tend to be oligopolistic, in common with other high technology industries, and it is predicted that ultimately this will be the case in most of the industry.

Further reading

Economics of the Pharmaceutical Industry. (1982). J. W. Egan,
    H. N. Higinbotham & J. F. Watson. New York: Praeger.
*The Chemical Economy. A Guide to the Technology and Economics of the
    Chemical Industry.* (1973). B. G. Reuben & M. L. Burstall. London:
    Longman.
The Chemical Industry. Social and Economic Aspects. (1972). F. R. Bradbury &
    B. G. Dutton. London: Butterworths.

Biotechnology and industry. (1981). P. Dunnill. *Chem. and Ind.*, **7**, 204–17.

Price Theory and its Uses. (1977). D. S. Watson & M. A. Holman. Boston: Houghton Mifflin & Co.

Chemical Marketing Reporter is a commercial newspaper of the chemical industry and a valuable source of US prices for many biotechnological as well as chemical commodities.

Chemical and Engineering News, Chemistry and Industry and *European Chemical News* are news magazines which cover the chemical industry and many biotechnological products. They give financial information and the results of individual companies.

3 Innovation and economic analysis of projects

Introduction

Biotechnology is widely perceived to have a high rate of innovation and to be highly research orientated. It has the potential to make many new products and it can provide alternative routes to existing products, but only a limited number of processes are actually in operation. It is important to be able to select and evaluate new projects in economic terms. It is therefore the aim of this chapter to tie biotechnology in with the general theory of technological developments and to outline the economic and accounting principles which apply in the evaluation of new projects. This can also provide insights into the strengths and weaknesses of the technology, highlight cost sensitive areas and indicate where improvements need to be made.

Technology and product life cycle

Many products, processes and technologies are frequently described by a theory known as the product or technology life cycle in which their development can be split into definable stages (Fig. 3.1). According to this theory sales and profits from a product, process or technology are determined largely by the stage of the cycle reached. In the early stages of conception, development and applications testing there are no earnings, only expenditures; there is thus a negative cash flow. In the launch stage there is heavy expenditure on promotion for a product or commissioning for a process. Profits are only realized after a successful launch in a growth phase. Increasing sales and technology improvement reduce unit costs. As profits increase imitators are attracted, market demand becomes satiated, costs cannot be reduced further, and the maturity phase is reached. Following this a decline sets in, with falling profits and sales as new products, improved technologies and processes erode the established position.

The validity of this concept is often challenged because many products do not appear to decline, or the maturity phase is maintained for so long that the cycle model has only a limited value. For example, many pharmaceuticals have a short life span, being supplanted by more effective replacements, often in less than 20 years. Antibiotics, however,

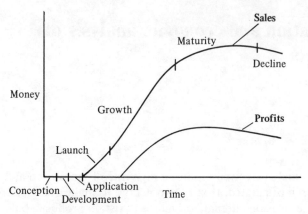

Fig. 3.1. Technology or product life cycles.

have been one of the longer lived groups. Penicillin has been a successful antibiotic for over 40 years and its market share is still strong. Its production levels have gone on increasing because it is now the raw material for a whole range of semi-synthetic derivatives. Other products of biotechnology have had shorter lives; for example production of acetone and butanol by fermentation was only pre-eminent for 20 years before being supplanted by petrochemicals. Glucose syrups and now perhaps high fructose syrups are entering the maturity phase, but who can predict how long it will last? Sucrose has had a maturity phase approaching 100 years in industrial countries. Many products over 100 years old have never been completely replaced by substitutes.

Given these limitations this model does emphasize the importance of constant product and process innovation and it is particularly relevant in biotechnology today. Monoclonal antibodies may replace radioisotopes in many clinical assays, but present monoclonal methods may be short-lived because of the rapid developments in this area. Two factors are important here: market pull, as in the case of antibiotics, and technology push, as represented by recombinant DNA and monoclonal antibodies. Both influence the decision to innovate.

When applied to a technology, the conception/development phase occurs when the major technical advances take place. It is the enabling phase. During the growth phase, investment and effort are directed towards exploiting these technical advances by improvements in production and process technology. Significant improvements may be made as experience accumulates on the learning curve (Vaughan, 1984). Eventually the ultimate performance limits of the technology are approached. As first proposed by Schumpeter (1934), there is a discontinuity between technologies. The development of each technology may be represented as in Fig. 3.1 with a clear break from its predecessors. It was best expressed by Schumpeter; 'Add successively as many mail coaches as you please, you will never get a railway thereby'. The pattern can be seen clearly in many historical developments, for example steam engines, aircraft or plastics. The chemical industry is thought by many to be in the maturity phase, but the cycle theory depends on the definition of a technology. Chemistry developed with coal as a feedstock, but then grew rapidly when oil was adopted. It may grow rapidly again with a new feedstock. Similarly with biotechnology, genetic engineering is in the innovative phase, fermentation is a mature technology, and new reactor designs may be required for a growth phase.

Innovation

Innovation is now widely recognized as being a cornerstone of the success of economic systems. At a national level a high rate of innovation is correlated with increasing output, productivity, employment and a favourable balance of trade. It is also seen as a weapon against inflation and a way out of recession. For example, information technology and microelectronics are viewed as a counter to the recession of the early 1980s. At the level of the individual firm, innovation is essential both to prevent overhaul by competitors and for long term survival. Because innovation is playing a key role in biotechnology today it is worth describing the process in more detail.

Scherer (1970) has divided the process of innovation into four separate functions. He defines *invention* as an act of insight by which a new technical possibility is recognized and worked out in its essential and elementary form. It is followed by *development*, a sequence of testing and modification to prepare the original concept for commercial exploitation. The *entrepreneurial* function is concerned with organization and the acquisition of financial support, while the fourth component, *investment*, is the act of risking funds for the venture. It is very informative to analyse the new ventures in biotechnology (Chapter 11) in these terms.

The ability of a company to produce significant innovations is related to the amount it spends on research and development. There is also evidence to suggest that a company's rate of productivity increase is also related to its research expenditure, but with a time lag. There is considerable industry variation in research expenditure when expressed as a proportion of total sales (Table 3.1). In general the industries characterized by high volume, low added value products spend the lowest proportion. The highest levels are found in those areas where product novelty rather than price has become the major factor in competition, notably pharmaceuticals and defence. Biotechnology spans

Table 3.1. *Expenditure on research and development as a proportion of gross sales, industry by industry*

Industry	Percentage	Source
Food and kindred products	0.4	National Science Foundation
Petroleum refining and related industries	0.6	National Science Foundation
Industrial chemicals	3.3	National Science Foundation
Computers, peripherals, telecoms	5–6	McShane (1984)[a]
Pesticides (chemical)	7	
Aerospace, defence	6–12	McShane (1984)
Pharmaceuticals	12	Rapoport (1983)[b]
Individual enzyme and pharmaceutical manufacturer	15	Gist Brocades Annual Report (1982)

[a] O. McShane (1984). Commercialising an extremely thermophilic ecosystem. In *The Proceedings of Biotech '84 Europe*, p. 150. Pinner: Online Publications Ltd.
[b] *Financial Times*, 15 November, 1983.

the full range from food to pharmaceuticals, so it is not quite accurate to describe it as a whole as having a high rate of innovation. For some sectors this is so, but biotechnology also serves other industries whose relative rate of spending on research is low. Similarly the rate of innovation in the traditional fermentation industries has been low, but the new companies will be in the upper echelons of the research expenditure league.

An important new concept in high technology industries is that of technology transfer, or bridging the gap between universities where the inventions have been made – funded by government – and exploitation of the inventions by businesses – the entrepreneurial and development functions. This topic has been well reviewed by Vaughan (1984). She targets three major obstacles to technology transfer: poor communications, differing priorities (e.g. publication, patenting, different career development) and inappropriate administrative systems. Many systems are being evolved in different countries to overcome these difficulties and facilitate the process. They include consortia, spin-off companies, science parks on or near university campuses, and industrial liaison groups. The United States has achieved the greatest success in this area through both its venture capital system (Chapter 11) and offering places on the boards of new companies, with direct financial incentives and stock, to inventors and academics.

The role of large and small companies in the innovative process is still a matter for debate. One school of thought maintains that large companies with monopoly power can sustain a greater rate of innovation than small companies in a competitive situation. This is

because the large expenditure on research and development can only be
generated by companies with large cash flows and large margins
protected by a monopolistic trading position (Schumpeter, 1942). This
model with monopolies being granted for a limited time by patents has
been effective in stimulating a high level of innovation in
pharmaceuticals. On the other hand small venture companies in the
clinical/diagnostic field have also sustained an impressive rate of
introduction of new products in an atmosphere of intense competition.
Some estimates put small companies as being 2–4 times more
productive in innovations per employee than large companies and
obtaining many times as many innovations per unit of research
expenditure as large companies. The answer may lie in the nature of the
product. Many products with limited markets are not attractive to large
companies with high overheads. Small companies can generate sufficient
returns from low sales volumes, if coupled with high margins, to
produce new and improved products, provided the margins can be
sustained by constant innovation. In areas such as food and
pharmaceuticals, regulatory costs alone preclude all but large
organizations operating on a world-wide market unless a link, such as a
limited partnership, is made between a small innovator and a large
company.

The least quantifiable character of innovation is uncertainty or risk.
For example it has been claimed that 3000 candidate compounds are
examined for every one effective drug produced by the pharmaceutical
industry (Davies, 1967). Some 5500 antibiotics had been described by
1981 and they are still being discovered at the rate of about 300 a year,
yet only approximately 100 products are on the market (Demain, 1981).

In common with most industrial research, biotechnology has been
dominated by empiricism. This does stress the fact that effective
research and development hinges not only on the ability to generate
useful inventions, but also on the timely recognition of losers and their
elimination.

Patents and licensing of inventions

A patent is a right granted by a government agency to an inventor
giving him an exclusive right to make, use or sell his invention. It is a
form of deal whereby the inventor must make a public disclosure of the
invention in return for being granted monopoly status with respect to it
for a fixed term. Patenting is necessary to ensure that producers of new
inventions or innovations receive a return on their investment in
research and development. It is justified as being essential to induce
innovation and to support research. Information may be expensive to
produce but relatively cheap to copy. In biotechnology as elsewhere

patents are an indispensible element in research and development, and much effort must be directed to ensure that work is patentable, otherwise it may have little commercial value. The importance of patent law to biotechnology is covered in detail elsewhere (Crespi, 1982; Perry, 1984), but it is included here in outline because of its significance.

For work to be deemed patentable it must be demonstrated to have novelty, inventiveness and utility. Ideas or theories cannot be patented; there must be a practical application. The invention must not be obvious to those skilled in the art, that is it must not follow on simply and logically from what is already known about the subject, what is described in legal terms as the prior art. Obviousness is a subjective concept and difficult to define, but it can sometimes be overcome by claiming difficulty or unexpectedness. All forms of public disclosure of the invention contribute to the prior art. This need not be confined to the literature or the press; verbal disclosures or commercial use which is public, even if made by the inventor himself, will invalidate a patent claim. In the US an inventor is allowed 1 year in which to file a patent application after public disclosure.

A patent claim is first filed with a patent office, usually in the country of origin, and its date of receipt established as the priority date. Within 12 months of this date the inventor must submit a full application with all supporting experimental details, but the claims must remain unaltered. Also, usually within 12 months of the filing date, the application must be submitted to the patent offices of all the countries for which cover is required. Europe (up to ten countries) can now be covered in one overall application. The application is then examined by the patent office and published if successful. If no objections are raised by others upon publication it is generally granted within three months. Patent cover now extends for 20 years from the date of application in most countries, although it is even less in some. It is 17 years in the US and Canada, but this is calculated from the grant of the patent.

Patents which claim a new microorganism or mutant must be accompanied by the deposit of that organism in a recognized culture collection before the date of filing. The organism must be accompanied by a detailed description of its physiology and morphology and this information must also be included in the patent application. The organism need now only be deposited in one culture collection if that collection is recognized under the Budapest Treaty of 1977. The depositor may restrict the release of the organism to third parties until the patent is granted, but from that date it is public property and available to all, although the depositor may be notified of recipients.

A patent contains two important parts; the description of the invention, including details of how the experimental work was performed, and the claims. The description must be sufficiently detailed so that the work can be carried out by a skilled individual and the results reproduced. The claims are aimed to be as broad as possible, but they must be supported by the experimental examples. Patents are a mixture of science and law, and are generally drawn up by patent agents who are able to understand the invention and convert it into the legal language necessary to gain the maximum protection.

Patent costs and agents fees begin at around $1000 to file an application and another *c.* $500 to deposit an organism. One year later it costs at least $1000 per country to file, including costs for translations and agents in each country. There may be further charges in supporting the application followed by maintenance fees. Some countries demand these annually at an increasing fee with time, while some demand them less frequently and some not at all. It is not possible to give an accurate overall cost because it is dependent on the number of countries covered and the support required, but an approximate estimate for cover in 20 countries is in the region of $50000 over the life of the patent. Ownership of a patent is ascribed to those who funded the research activities, i.e. usually a university, a government agency or a company. These may have policies of remunerating staff but the legal ownership is theirs. In the case of outstanding inventions the employee inventor may have the right to seek compensation in some countries.

It is possible to patent products (including microorganisms, enzymes, plasmids and cell lines), processes (including production methods, special techniques and diagnostic methods), compositions or formulations and new uses of an existing product. Some countries do not allow pharmaceutical compounds to be patented, but they are now few in number and do not represent particularly significant markets. An existing organism can be patented in a new process or for a new function, and a host organism containing new genetic information, such as plasmids or other vectors, and now DNA and RNA sequences, may also be claimed. In Europe microbiological processes and products are patentable, but plant or animal varieties are not, although a new vector or nucleotide sequence from a plant or animal could be claimed.

Microorganisms are now patentable in the US, but this development is relatively recent, stemming from the landmark case of General Electric (Chakrabarty) in 1980, who finally successfully claimed a strain of *Pseudomonas* containing a number of catabolic plasmids able to decompose oil slicks. Other notable recent patents in biotechnology are those granted to Stanford University (Cohen and Boyer). The first covers the recombinant DNA process in unicellular organisms and the second covers products made by recombinant DNA techniques. If the techniques are used for research leading to a commercial product the user must obtain a licence from Stanford. The products are deemed to be commercial when a first application is made to the US Food and

Drugs Administration for approval. Academic research is excluded
from the licensing requirement. The cost of a licence is $10000 p.a. plus
a royalty on sales of any product made by the technique of 1% of the
first $5 million in sales ranging down to 0.5% of sales over $10 million.

Biotechnology now enjoys widespread patent cover following a
number of key decisions in recent years. Some biotechnology companies
are prolific patenters. For example it has been reported that Genentech
has about 1400 patent applications outstanding. In general, patents are
important to the commercial exploitation of inventions in biotechnology
and will stimulate innovation as elsewhere, but they are not without
their drawbacks. Obviously the cost of the patenting processes must be
justified by the potential value of the invention; however, the ultimate
value is not known, and many or most patents are never used – i.e.
there is an element of risk. There is also the factor of the publication of
detailed information which may be of benefit to competitors.
Sometimes a decision may be made to keep the information secret
rather than seek a patent. Also with care and effort it is often possible
to design around a patent. Process patents are generally more
vulnerable here and are therefore usually of less value than a product
or use patent. Many companies may spend much money on research to
get around patents by minor modifications. In the food and drugs
industries, however, the competing product will also have to be cleared
through a regulatory authority, the costs of which dwarf the patent
expenditure, so in general patents tend to be more valuable in these
industries. On the other hand the time taken to get the regulatory
approval may be a significant proportion of the time covered by the
patent. For example Aspartame, manufactured by G. D. Searle, came
onto the market in 1983, but in many countries its patent expires in
1987.

The licensing of inventions is a way of selling technology which can
be a quick route to recovering research and development expenses. Its
application to biotechnology products is reviewed by Wheaton (1984).
A legal agreement is drawn up in which royalty rates and other
conditions are determined. The royalty is usually a percentage of the
licensee's sale price and less frequently a fixed sum, but there may be
minimum royalties payable. Royalty rates are dependent upon
exclusivity and the geographical area covered. Often one manufacturer
may sell a licence because of a lack of sales or service networks in
another country. Successful licensing is highly dependent on trust,
because litigation is invariably expensive. Standards of performance of
the new technology must be set and agreement reached on the efforts
that the licensee should make to manufacture and sell the product.
Both are contentious areas and may require training and assistance by
the licenser.

The background to economic analysis of projects

Economic planning is concerned with the allocation of scarce resources to gain maximum benefits. This has both macro- and microeconomic dimensions which affect biotechnology. On a macroeconomic level the resources of a country such as labour, capital, land and foreign exchange can be applied to a variety of uses such as production of consumer goods, public services, food or fuel. This type of planning has been undertaken by the United States, Brazil and India with reference to ethanol production, and is described in Chapter 8. Other nations are actively considering similar programmes concerning diversion of agricultural resources towards fuel and feedstocks. On a microeconomic level planning relates to decisions by individual firms to develop biotechnology as a means of production and to support individual projects which are considered to represent the maximum benefit from available capital, labour, plant or feedstocks. The ultimate aim may be to maximize profit, but many other factors must be considered on the way, cash flow and break-even point in particular. Many bankruptcies are caused by firms becoming overstretched, even though the project is basically sound and can eventually be profitable. In essence project analysis measures the costs and benefits of a project in common terms. If benefits exceed costs development is continued; if not the project should be rejected.

Forward planning perhaps represents one of the most fundamental differences between academic and industrial research. It is also one of the most contentious areas within industrial research. The differences stem ultimately from the perceived benefits. Academic research is concerned with long term benefits and with all areas of society. Industrial research is concerned with the individual firm, usually over a much shorter timespan and with scale-up and process development always in mind. It must also carry all its costs including research, whereas academic research generally involves a high degree of cross-subsidy, mostly from education budgets in terms of buildings, infrastructure, some salaries and so on. Corporate management of research must also consider the firm's expertise, the fluctuations in the company's fortunes and its existing markets. Marketing considerations are paramount; numerous case studies of successful and unsuccessful commercial projects point to a strong positive link between marketing, research and development.

The scarcity of resources applies at all levels through basic research to plant construction, and costs increase exponentially with time and scale-up as represented in Fig. 3.2. As an approximate rule of thumb in biotechnology, pilot plant development costs 10 times more than laboratory work and full scale plant costs are at least 10 times more

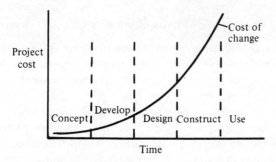

Fig. 3.2. The cost of change. From Kharbanda & Stallworthy (1982).

than pilot plant work. All projects are thus subjected to a continual process of financial appraisal. The increasing costs must be matched by increasing sophistication of appraisal. At first the analysis is necessarily crude because of insufficient data, but development work must provide data which permit ever more accurate costings. The analysis must cover all costs, that is those incurred by research, development, capital, production, marketing, distribution and so on.

In addition to this straightforward reasoning a study of planning methodology is also valuable for the insight it permits on the strengths and weaknesses of individual processes, methods of production and entire industries or technologies. This analysis often throws up completely unexpected results. The usual elements overlooked or underestimated are capital costs and the time element involved in their repayment, research costs (increasingly the costs of obtaining regulatory approval) and other overheads as opposed to direct production costs.

The evaluation of these costs and benefits is difficult enough in direct terms to firms in a capitalist economy, but there are accepted techniques, ground rules and standards, some of which are described in this chapter. The evaluation on a national level and long term strategic planning by both large corporations and nations is fraught with difficulties and loaded with value judgements. Some issues will be described at various points in this book, but a systematic coverage has not been attempted. In general the contentious issues with relevance to biotechnology which lie outside the direct planning methods described here are:

(1)   Environmental.
(2)   Political – employment, import substitution, balance of payments.
(3)   Long term technological capability – 'seed corn'.

Table 3.2. *Project appraisal checklist*

	Score:	V. poor	Poor	Average	Good	Excellent
	Value:	1	2	3	4	5

1. Technical factors
   A Basic principles
   B Company expertise
   C Success of other researchers
   D Chance of Co.'s R & D success
      compared with others
   E Compatibility with other
      parts of R & D programme
   F R & D timescale
   G Can company perform
      applications testing?

2. Commercial Factors
   H How many areas of application?
   I Customer price advantage over
      competitors
   J Value to customer advantage
      over competitors
   K Market size
   L Market trends
   M Competing products
   N Company proprietary position
   O Fit with company's planned
      areas of activity
   P Company's commercial expertise
      in field
   Q Capital expenditure required
   R Market timing

Project selection

Many potential project ideas can be screened out by very approximate calculations based on the prices of starting materials and the prices plus sales volumes of competing projects. For example microbial lipids must have properties which would give them values several fold higher than the oils obtained from plants such as palm, sunflower, rape or corn because the substrates for growth of the microbes are in the same price range as the oils (Table 4.3). Beyond this, selection of projects must be put on a more objective basis weighing factors such as the firm's expertise, its current markets, development times and so on. A sample project appraisal checklist is shown in Table 3.2. Each factor is assigned a score value on a scale of 1 to 5, 1 being very poor, 3 average, 5 excellent. For example the time scale might be rated 1 for over 10 years,

2 for 5–10 years, 3 for 3–5 years, 4 for 1–3 years, 5 under 1 year. The criteria for selecting candidates for further evaluation in this system are:

> $I+J$ greater than 5
> $K+L$ greater than 5
> $(M+N)\times(O+P)$ greater than 40
> $K\times Q$ greater than 4
> $B/(A+C)$ greater than 0.3
> $D-C$ must be positive
> $F\times R$ greater than 7
> $(M+N)\times(K+L)$ greater than 40
> Sum of $A–R$ greater than 50.

This method is clearly arbitrary, but in fact it only screens out grossly unsuitable projects. The actual method used varies from company to company, but it is important simply to illustrate the criteria involved and their approximate weightings.

Evaluation of projects is made in economic terms, but not just in money terms. The effect on existing company assets, sales range, technical skill, commercial expertise and manufacturing skills is of great importance. Synergy or special advantage are sought. Synergy is a big advantage in highly competitive markets, and thus oil companies have tended to develop forward integration into chemicals rather than say diversification into soft drinks (dissynergy).

Research and development planning

The progress of research and development projects within a firm can be represented graphically in the form of a decision tree (Fig. 3.3). From each node two or three branches lead to other nodes. Each fork is either a decision fork, where the branching represents choice on the part of the planner, or it is an event fork representing factors over which the planner has no control. Examples of events which might be relevant are construction of a similar plant by a rival firm, price changes in raw materials or products, currency fluctuations, failure of toxicology tests or changes in government policy. This type of plot is a graphical illustration of progress but it is not particularly valuable in planning. Various methods have been developed to co-ordinate research activities with one another and to plan development of projects with time. Many of the techniques have been developed to present planning in a visual form. The simplest of these is the bar chart. This consists of a horizontal bar to represent each activity in the project. The length is proportional to the time it will take. An example of a bar chart for a completed project, the Rank Hovis McDougall Mycoprotein project, is

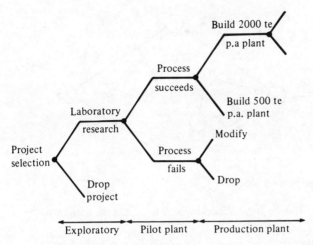

Fig. 3.3. Decision tree in research and development management.

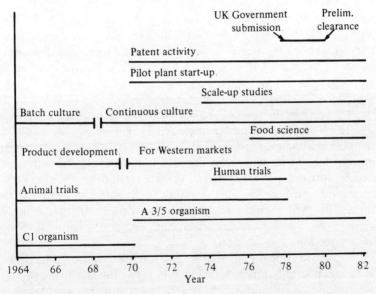

Fig. 3.4. Twenty-year research and development cycle for mycoprotein.
Reproduced with permission from Dr J. Edelman, RHM Research Ltd.

shown in Fig. 3.4. If such a chart were examined in the course of the project's development it would show completed sections perhaps as a shaded bar and planned operations as an open one. It illustrates the timespan of major projects and the way different activities such as pilot plant trials, human food testing, patent claims etc. must be co-ordinated and run in parallel. The technique can show which parts

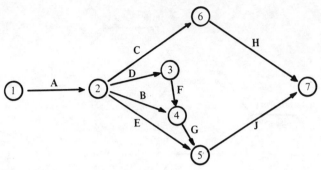

Fig. 3.5. Project network analysis.

of a project are ahead or behind schedule and where resources must be re-allocated.

Bar charts, however, only show activities on a time related base, they do not show the interdependence of various activities, or give a guide as to what effect a delay in one activity will have on the total project. This can be achieved by network analysis (Fig. 3.5) where events (numbered circles) are linked by activities (arrows). Events are the start of the carrying out of a particular action, for example fermenter delivered, fermenter installed, fermenter operates. Activities are the elements of the project, for example install fermenter, or write computer programme. Networks can show the interdependence of activities. In Fig. 3.5 event 4 (the start of activity **G**) cannot take place until activities **F** and **B** have been completed. They have the advantage in that adding the times of activities on each route between the start and finish of the project will reveal the rate-limiting or critical path for the project. Any delay in this series of activities will delay the completion of the project unless action to make up the deficiency can be taken. They can also be used to calculate the float time for an activity not on the critical path, that is the delay which can be tolerated before it affects the completion of the project.

Network or critical path analysis is a complex subject in its own right, and it is only briefly mentioned here to illustrate the interrelationship of events in project development. For small projects a network can be constructed by hand, but larger projects require computers. Unlike in the bar chart method the current status of a project cannot be shown in a network diagram and any alterations require a reconstruction of the whole chart, which in manual operation is a tedious task. Some of the advantages of bar charts and network analysis can be combined in a technique called Job Progress Charting in which the horizontal lines in bar charts are linked by vertical lines where th∙ ∍nd of one activity represents the commencement of another, and slack time can be represented by dotted lines (Fig. 3.6). In this case

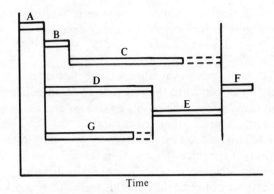

Fig. 3.6. Job progress charting.

A–D–E–F represents the critical path. In a typical example **A** might represent basic research, **B** production of product for toxicological testing, **C** toxicology, **D** pilot plant trials, **E** construction of full scale plant, **F** marketing, and **G** market analysis.

Capital budgeting

Biotechnological processes require specialized equipment engineered to fine tolerances in reactors and often in product recovery. Also in general plants must operate aseptically, they need analytical equipment, effluent treatment and increasingly sophisticated control equipment. They are therefore highly capital intensive. The recovery of capital costs is one of the most significant factors in project analysis. Several methods of varying complexity can be employed. They are all covered in accounting texts (see Further reading) but are worth describing briefly because of their importance and their bearing on many otherwise attractive projects.

Payback period

This is simply the time needed to recoup the initial money invested, in the form of net cash flow from operations. For example, assume that a reverse osmosis apparatus is purchased for $20000 and it saves steam costs of $5000 p.a. in evaporators to achieve the same amount of dewatering. If the electricity, membrane and other operating costs of the reverse osmosis apparatus are $1000 p.a. the payback period is $20000/(\$5000-\$1000) = 5$ years. It is a rough and ready model, which does not measure profitability. It does not mean that a project should be selected in preference to an alternative project with a longer payback period; for example, the useful life of alternative pieces of equipment must be considered. Nevertheless it is widely used and quoted, usually on equipment to modify a process or in biotechnology in unit recovery operations such as centrifugation or drying.

Return on investment

A first step in the financial evaluation of a project is to calculate the ratio of annual profits to the original capital investment. There are two important variables: the return on capital and the price of the product. If the price can be estimated accurately the return on capital can be calculated and a decision made on the favourability of the project. Alternatively if the price is not known the firm can dictate a desired return based on other projects or current interest rates (i.e. alternative applications of funds) and use this to calculate a price. If this price fits with the expectations of the market the project can be approved.

This can be illustrated in a plan to build a small plant to manufacture 300 te per year of fermentation product such as a microbial surfactant, yeast biomass or a microbial insecticide which have simple downstream steps. The process involves the use of 2×20 m³ fermenters operated in batch mode out of phase. Recovery of the product is by centrifugation of the broth followed by formulation and spray drying to obtain a fine particle, dry, wettable powder. The capital and operating costs for this proposal are shown in Table 3.3. They show that to achieve a return on investment (ROI) (internal rate of return) of 20% p.a. the product must sell for £4165 te⁻¹. If the rate were set higher at 30% then a price of £4802 te⁻¹ must be obtained. At the time this exercise was done the selling price of the product was £4250 te⁻¹ (21% ROI), but this calculation takes no account of research and development costs. The exercise shows that on small projects like this the capital charges are very important. The variable costs such as raw materials and utilities only account for 28% of the selling price.

This example also illustrates the importance of the rate of return on investment demanded by the company, which in general is determined by risk and is very industry dependent. Large corporations used to dealing with high volume, low margin products such as basic foods and chemicals have lower expectations of return than pharmaceuticals with low volume, high added value and high risk ventures. The rate also reflects both the general level of profitability on turnover of any industry and the other company overheads. Pharmaceutical companies in general are highly profitable and they require contributions from individual projects to fund their extensive research programmes.

Discounted cash flow

The method of calculating return on investment can be used as a guide in feasibility studies, but it does not allow for the time factor of

Table 3.3. *Cost schedules for a small fermentation process*
(*300 te per year plant*)

Capital costs	£
Plant and equipment	550000
Spares (at 5%)	27000
Buildings (fermenter house, formulation & drying area, lab. and warehouse)	100000
Installation costs (by Lang factor)	1100000
Commissioning	15000
Vehicles	20000
Development costs	100000
Total	1912000

Annual operating costs	£/te
Variable costs	1200
Production overheads, maintenance	400
Administrative expenses	133
Engineering charge	166
Management and sales	166
Contingency (20%)	825
Annual capital charge at 20% ROI	1275
Total	4165

a development. The movement of money with time, or cash flow, in a
typical example such as the fermentation project is shown in Fig. 3.7.
In the early stages of such a project research and development costs
produce a small debit which increases with pilot plant work and
production plant construction. It is only after the plant has been
commissioned and product is being marketed that a positive cash flow
is achieved. The area under the line represents total expenditure before
any sales are made and the area above the line net cash flow from sales,
i.e. the contribution from sales after deduction of operating costs, raw
materials and utilities (Table 3.3). In this example it has been assumed
that plant construction takes 2 years after 1 year of pilot plant work
and 2 years of research and development. These are all low estimates.
In the sixth year it is assumed that sales are only half their final annual
value and commissioning costs reduce net income. Two selling prices of
£4165 te^{-1} and £4802 te^{-1} (based on ROI of 20% and 30% respectively)
are shown. If the areas above and below the line are compared it can be
seen that the payback period (straight cash repayment) is 10 years after

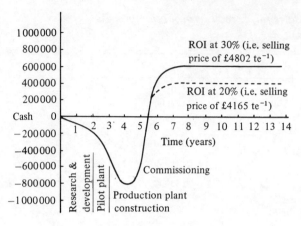

Fig. 3.7. Cash flow diagram for construction of 300 te p.a. fermentation plant.

the project's commencement or 7 years after the start of plant construction at the higher selling price. At the lower price these periods are extended to 12 years and 9 years.

This calculation does not take into account the alternative uses to which the money could have been put over this period, for example the receipt of interest if it were invested on money markets. This is most commonly corrected for by an accounting practice called discounting, and corrected discounted cash flow diagrams constructed. This technique allows for the time value of money, i.e. the sooner money is obtained from an investment the more valuable it is. It can be invested again. The discounting method uses factors for the minimum rate of return chosen so that the cash flow can be converted to a common base, termed the present value. For example a £100 income expected a year hence has a present value of $£100/(1+r)$ where r is the discount factor or appropriate interest rate. If the interest rate is 10% p.a., $r = 0.1$ and the present value of the £100 is £90.91 (i.e. if £90.91 were invested today at 10% p.a. interest rate it would be worth £100 in a year's time).

If this is applied to an investment which is predicted to yield profits before depreciation of P_1, $P_2 \ldots P_N$ in the 1st, 2nd $\ldots N$th years of its life, the present value is given by:

$$\frac{P_1}{(1+r)} + \frac{P_2}{(1+r)^2} \cdots \frac{P_N}{(1+r)^N}$$

$$= \sum_{n=1}^{n=N} \frac{P_n}{(1+r)^n}$$

The project must also repay its capital cost. If the depreciation payments are $D_1, D_2 \ldots D_N$, their present value is similarly given by:

$$\sum_{n=1}^{n=N} \frac{D_n}{(1+r)^n}$$

If taxation is ignored the cash flow C is the sum of profits and depreciation:

$$\sum_{n=1}^{n=N} \frac{P_n}{(1+r)^n} + \sum_{n=1}^{n=N} \frac{D_n}{(1+r)^n}$$

$$= \sum_{n=1}^{n=N} \frac{(P_n+D_n)}{(1+r)^n}$$

$$= \sum_{n=1}^{n=N} \frac{C_n}{(1+r)^n}$$

Two general methods are used. In the first the current rate of interest is given and the present values of the cash flows is estimated. For example approximate cash flows for the fermentation project for 15 years are shown in Table 3.4. To simplify the mathematics, a selling price of £4223 te^{-1} has been selected so that when operating costs of £2890 te^{-1} (the capital charge of £1275 in Table 3.3 is not included) are subtracted there is a cash flow of £1333 te^{-1}. Then if the plant is operating at its maximum capacity of 300 te p.a. there is a cash flow of £400000 p.a. In year 6 capital expenditure is still occurring but income from product sales is starting; the net cash flow has been assumed to be zero. To simplify the mathematics further, an interest rate of 10% has been selected to give the present values. Using this method the return on the project is exactly equivalent to having invested the money in a bank at 10% interest during the 15th year; at the end of this year the project has a small positive net present value. It could therefore be considered desirable at this rate of interest and a plant life of 10 years after construction. If a higher rate of interest were desired or plant life were shorter it would be unfavourable. In practice unless prevailing interest rates were very low, a higher rate of return, probably at least 15–20%, would be required for a project of this type. When these figures are drawn on a plot the discounted cash flow diagram obtained is one of the most informative ways of displaying the financial implications of investment in a new project. It can be seen that the costs of research and development coming early are not heavily discounted and their effect on present value is relatively undiluted, whereas the income from the plant coming late suffers a heavy discount and contributes less to the balance.

Table 3.4. *Discounted cash flow for fermentation project*[a]

Year	Cash flow ($\times$£1000)	Discount factor at 10%	Present value ($\times$£1000)
1	−20	0.909	−18.1
2	−40	0.826	−33.04
3	−200	0.751	−150.2
4	−700	0.683	−478.1
5	−800	0.621	−496.8
6	0	0.564	0
7	+300	0.513	+153.9
8	+400	0.467	+186.8
9	+400	0.424	+169.6
10	+400	0.386	+154.4
11	+400	0.350	+140.0
12	+400	0.319	+127.6
13	+400	0.290	+116.0
14	+400	0.263	+105.2
15	+400	0.239	+95.6
		Net present value	72.86

[a]Assumes selling price of product of £4223 te^{-1}, and that income balances expenditure, i.e. net cash flow is 0 in the 6th year.

An alternative method is to use the discounting technique in a trial and error fashion to determine the interest rate r. A value of r exists such that the present value of the positive cash flow expected from the project just balances the expenditure. This value of r is termed the internal rate of return because it is determined only by the magnitudes of the cash flows and not by anything outside the project. The cash flows are tabulated before discounting from the start of construction to plant commissioning. Trial discounting rates are then applied to see which rate makes the present value of earnings equal to the present value of expenditure. In the fermentation example over the 15 year span this would be approximately 10%. If the prevailing interest rate is greater than this internal rate of return the project will lose money; if it is less it may be considered favourable, but then risk and other factors must be considered. This method has the advantage that is does not demand a specified minimum discounting rate, but it is tedious because it requires solving the rate by trial and error. Both methods are now generally calculated using computers or programmable calculators rather than the tables traditionally used.

Discounting is covered in all accounting texts, and is outlined here only to illustrate the importance of capital expenditure to the type of project encountered in biotechnology. For example it shows that the fermentation project, which at first glance in Table 3.3 might appear favourable, is not so unless prevailing interest rates are very low. It also makes definite conclusions about the life of a project. In this example the plant must operate at full capacity for 10 years without any further expenditure to achieve a rate of return of 10%. It is also valuable because expenditure and receipts are listed as they are expected to occur. Expenditure on plant construction or modification is not usually a single down payment, but is spread over years depending on the terms of the contract. Plant commissioning costs are an important unknown and may dramatically affect the viability of a project. As in the return on investment method, discount rates depend on the company and perceived risk. Rates are generally chosen between 10 and 20%, but projects of an uncertain nature may be discounted at say 25% and safe projects only marginally higher than the prevailing borrowing rate.

Though discounting is an elegant technique its construction makes great demands on detailed information or estimates of expenditure and income over the project's predicted life. The calculations can be done accurately but there is always uncertainty with the input data. The procedure as described does not take account of inflation. Constant inflation means the real value of a given sum at a future date will be less than the present value, but income from the project in actual terms will be higher. Also in this and other budgeting techniques no account of taxation has been considered. This can transform the economics of many projects as capital expenditure may be allowed directly against profits and in some countries may be allowed in its entirety in the first year.

Costing

The object of costing is to analyse the expenses and income of a project. It determines to which product or service a particular expense or income relates and provides information for planning and the determination of product pricing. Two methods are generally used – full costing and marginal costing.

Full costing is an analysis of all the expenses of a business or project which allocates them to particular products or services. Costs are either direct, in which case they can be attributed to the product or service alone, or indirect where they must be spread over more than one product or service. Indirect costs are commonly referred to as overheads. Direct costs in biotechnological processes include raw materials, processing chemicals, utilities used in the reaction or product recovery, labour of process workers and plant depreciation. Indirect costs include land factors (rent and rates), amenities, offices and administration, heating and lighting. Full costing is primarily concerned

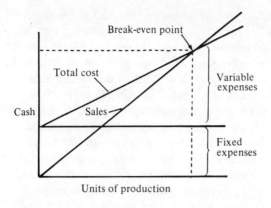

or:

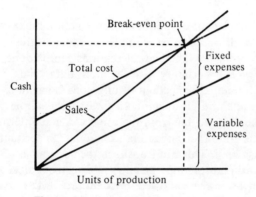

Fig. 3.8. Marginal costing.

with detailed analysis of plant operation. Examples for fermentation processes are shown in Chapter 5. In an appraisal exercise on a fermentation plant a full costing requires detailed prices of fermenters and driers of different capacities, pipework, packaging equipment and so on. It is an expensive procedure in itself and not warranted until a later stage of project appraisal, although some assumptions on capacity and scale can be made and will be described.

Marginal costing is a simpler technique which is of value in planning exercises. It analyses expense and income in terms of variable costs which change with the volume of production and fixed costs which do not. Variable costs in biotechnology include raw materials, processing chemicals, energy costs and overtime. Fixed costs include depreciation, loans, insurance, plant overheads and most labour. Marginal costing is frequently represented graphically (Fig. 3.8). It has limitations in that costs change with time and break-even points will be affected. It is also difficult to draw a clear line between fixed and variable costs. For example labour is generally considered to be a fixed cost in process plants, since trained staff are not generally laid off during time of soft production demand. If however output falls or is below target for a longer period of time, process workers are generally laid off first, followed eventually by administrative staff, salesmen and research workers. A factor becomes variable as soon as a decision has to be taken as to whether or not it shall be replaced.

Given these limitations marginal costing is very useful in project planning. It enables definition of a price policy and the establishment of a relationship between volume of production and price. The variable cost also acts as a bench mark denoting the minimum sales price above which a contribution is made to the fixed costs which are incurred whatever is sold or produced. In the longer term all costs of production, fixed and variable, must be covered, but in the short term fixed costs remain even if there is no production. Only variable costs are saved by ceasing to produce and so, providing these are covered by receipts, production may continue and any surplus will contribute to fixed costs. This situation of marginal costing is frequently resorted to during periods of reduced demand, and is a common practice in the chemical industry. In recent years the drop in the world price of sucrose has sometimes forced manufacturers of high fructose corn syrups to operate plants on this basis though they are generally protected by tariffs on imported sugar (Chapter 9). It is a risky technique in terms of long term survival. Many manufacturers of fermentation products, notably solvents, operated on a marginal basis as petrochemical prices fell in the 1940s and 1950s, but they were eventually forced out of business as worn out plant could not be replaced.

Marginal costing can also be used to attack competition by fixing a sales price below that of competitors. This is often used to discourage entry of new competitors into a field or to squeeze small competitors.

Costs in biotechnology

Most cost figures used in this book can only be used as approximate guides to give some impression of how a process can be analysed in financial terms, for a number of reasons. For instance, most refer to proposed processes or emanate from consultancy firms, universities or research institutes. Manufacturers are generally reluctant to reveal actual plant operating costs because of supplying information to their competitors. Projected costs are generally used to highlight claimed advantages of one process over another and naturally tend to be optimistic. It is also difficult to compare costs of plants in different geographical locations because local feedstock costs may vary, grants or tax reliefs may operate, labour, transport and property taxes may be different and so on. There are also almost as many different ways of presenting costs as there are companies. It is often very difficult to make comparisons because the analytical breakdown and presentation of costs may differ. For example labour and utilities may be quoted for the whole process or they may be split and assigned to fermentation or recovery. Without having a very detailed cost breakdown, which is rarely available, comparisons are limited, but some conclusions can be drawn and some insight into processes gained.

Raw materials

Raw materials form a significant cost input in all biotechnological processes. In many cases they are the dominant costs, sometimes contributing up to 70% of the total costs of producing low value commodities such as single cell protein or ethanol. They can usually be clearly defined in most manufacturers' published costs so that direct comparisons can be drawn, though there are notable problems in the case of by-products.

Some important factors concerning raw materials must be considered when preparing costing analysis of a new product or process. Firstly, different prices prevail for different volumes of purchase; tank car loads or truck loads are less costly than smaller quantities such as drums. Large scale bulk supplies must generally be secured by forward purchase, and a special contract price must be negotiated, which may be lower than quoted market prices but which commits the buyer to purchase by the terms of the contract. Prices quoted in sources such as *Chemical Marketing Reporter* may sometimes refer to this type of contract. Freight costs are cited in published prices in several different ways. *Delivered* means that the seller pays the freight independent of location, *freight allowed* means essentially the same thing, *f.o.b.* (free on board) the producing plant or shipping points means that the buyer must pay transport costs to his own location. In bulk commodities these extra changes can make considerable differences to the viability of a process. The form of raw material is also important. Some products are sold as syrups or slurries, others as solids, all with varying degrees of purity. In calculating yields etc. raw materials must be analysed by their relevant criteria, for example, with molasses the amount of fermentable sugar must be considered on a weight basis.

Utilities

This term generally embraces electric power, steam, cooling water, compressed air, de-ionized water and refrigeration, though some companies may put them in different categories in a cost sheet. Utility costs vary around the world depending on countries' individual energy prices, but in general steam and electricity costs are much lower in

North America than in Europe, often by as much as 50%. They will also be lower in oil producing countries. Steam and electrical power are the major costs affecting fermentation processes, particularly for sterilization, mass transfer and product recovery. Cooling water is however a vital commodity, and may prevent development of plants where it is not available on a year-round basis to maintain reactor vessels at the required temperature.

Overheads

Overheads are spread over a number of products and operations and therefore must be apportioned to different projects or processes by some formula or allocated basis, which again will vary from operator to operator. In planning exercises they are the most difficult items to estimate accurately and are generally charged on a very loose basis, or as a percentage of plant costs.

It is important to remember that different companies will have different overheads. In general, large companies have large fixed costs, and calculations will eliminate many small projects on the basis of overheads alone. Smaller firms are more flexible, have lower fixed costs and can thus contemplate smaller projects. Overheads will also vary on whether a new green field site is being chosen or whether the project can fit into an existing factory complex.

Selling expenses

These costs include the salaries of salesmen, advertising in journals, advertising literature and a proportion of the costs in maintaining sales departments or offices. Sales costs can vary greatly across the range of products obtained by biotechnological processes. They are probably highest in pharmaceuticals where they can amount to as much as 40% of the selling price. At the other extreme they are low, perhaps 1% or less, in the supply of bulk ethanol. There are often considerable sales expenses in the launch of new products which must gain consumer acceptance, for example Aspartame has been supported by national television and magazine advertising.

Capital costs

Preliminary capital cost estimates of installed plant can be made from the delivered equipment cost by a technique known as ratio estimating. This is a rapid technique which gives approximate answers. It is better than outright guesses and is cheaper than drawing up preliminary plans. It was first developed by H. J. Lang (1948) using factors which can be multiplied against equipment costs to get completed plant costs. Lang estimated from a survey of chemical plants that:

$$\text{Delivered equipment} \times \begin{cases} 3.10 \text{ for solid process plants} \\ 3.63 \text{ for solid/fluid plants} \\ 4.74 \text{ for fluid process plants} \end{cases} = \text{Total estimated plant cost.}$$

These costs include construction and construction overheads. The total installed cost of process equipment and piping, including foundations, supports, vents and insulations, is very nearly independent of plant size and type. It is approximately 1.5 times the delivered cost.

In practice the Lang factor for fluid processing plants has been found to vary from 3.85 to 5.05, so the method has been refined by breaking down the costs of a unit into several categories for the purpose of preparing a preliminary capital estimate (Bach, 1960). For example a plant may be split into storage facilities, cooling towers, distillation equipment and pilot plant and a different ratio used for each. Hand (1958) quotes different Lang factors for different types of equipment: 4.0 for pumps, pressure vessels and fractionating columns, 3.5 for heat exchangers, and 2.5 for compressors. As there are so many similarities in fermentation or immobilized enzyme plants it should be possible to perform accurate ratio estimations if figures for present plants can be obtained.

Capital outlay in plant must be recovered in depreciation costs. There are two charges on the use of capital, the repayment and interest payments. In costing a plant the depreciation may be spread in equal instalments over the expected life of the plant, so that if this is put at 10 years, 10% of the capital cost is charged annually to the product. Alternatively a series of gradually decreasing instalments may be calculated.

There is some controversy over the life expectancy of biotechnological plant and because many processes are capital intensive this can be crucial in determining whether the go-ahead for a project is given. Chemical plant is generally written off over 8–10 years because process improvements occur at such a rate that the plant is often obsolete, if not life-expired, after this period. It has been claimed that biotechnological processes should follow these rules, but there are arguments that much fermentation equipment has a longer life span than this, say 15–20 years and, although downstream operations may change, the primary reaction vessels will remain the same. Thus only part of the plant should be written off over the shorter period. Different companies will probably apply different rules in this area.

Working capital, which is the sum tied up in raw materials, unsold product, unpaid accounts and cash assets minus current liabilities, also carries cost by way of interest charges. It is sometimes included in cost

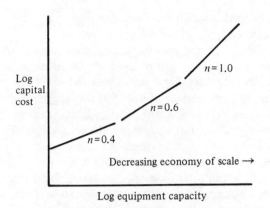

Fig. 3.9. Relationship between cost and capacity for high speed rotating equipment.

analysis figures and is calculated by relation to fixed capital or annual sales (Scott, 1978).

Economies of scale

Ratio estimations can also be used to relate the costs to the capacity of a plant. Collections of cost data for both individual equipment and whole plant construction costs have given rise to the so-called six-tenths factor. That is, if a plant size is doubled the cost will be $2^{0.6}$ higher, or alternatively a log–log plot of capital cost against capacity gives a straight line with a slope of 0.6. As with Lang factors this is used as a short cut in preliminary costings, but noticeable deviation is shown by many items of equipment. For example high speed rotating equipment such as centrifuges, pumps and fans show a lesser slope at smaller sizes, passing through a 0.6 slope in the intermediate range to a greater slope for larger sizes (Fig. 3.9). There is also evidence that some whole plants show a similar curve (Kharbanda & Stallworthy, 1982). Slopes greater than 1.0 indicate negative utility of size. As a consequence costs must be broken down according to types of equipment or installations to acquire more accurate estimates. In most economic appraisals capital requirements for a plant are estimated at at least two and more usually three levels of capacity.

Operating costs vary with scale according to the equation $X = yR^z$ where X is the unit cost and R is the production rate. Factor y must be determined for each cost item considered, for example raw materials, labour or utilities. Exponent z has been determined to lie in the range -0.30 to -0.50 for most industrial chemicals and is often assumed at -0.35 for preliminary estimates. As with capital costs there are individual variations and data from biotechnological operations is often insufficient to make accurate estimates.

The smallest economic production unit can be determined relatively reliably from cost sensitivity analysis, where the upper limit is determined by a variety of cost and non-cost factors. Of the latter, market forecasts are important and raise a problem with products having growing sales. If sales are increasing it is not wise to build only for the first year's expectations, but higher capital costs for larger plant have a negative effect on return. There are also physical limits to plant size such as engineering and construction know-how, the constraints of a site itself and transport factors. In addition there is the risk factor associated with cost overruns and commissioning delays which can be disastrous to a firm's cash flow and total finances. The larger the plant, the more expensive are the mistakes. Many other factors may be cited, such as organizational complexity and more expensive shutdowns. There is also a conflict between benefits to the manufacturer and benefits to the community as a whole in terms of large capital intensive plants *versus* small labour intensive ones. This is a controversial area. Large scale plants have won the day in the chemical industry with outputs often in the hundreds of thousand tons per year range. Small scale plants have been championed by E. F. Schumacher (1976), and some of the hazards of large scale developments are catalogued by Kharbanda & Stallworthy (1982). It is interesting to compare the considerations on scale which have shaped the development of fermentation ethanol plants in Brazil and the United States (Chapter 8).

Sensitivity analysis

This is a procedure which determines the sensitivity of the profitability of a project to changes in each of a number of factors. Sensitivity may be measured in terms of the variation of pay-back time, return on investment, net present value or internal rate of return against changes in costs which may include raw materials, utilities, yield, capital and so on. Taking the earlier fermentation project proposal, if the plant construction costs in fact turn out to be £2.5 million, not £1.9 million, the break-even point will be extended beyond 10 years, the return on investment will be lower than 30% at a selling price of £5000 te^{-1} and the internal rate of return will drop below 10%. The results may be tabulated in terms of upper and lower limits: thus if capital costs are put in a range of £1.9–2.5 million, pay-back period can be expressed as 10–13 years, while if selling price is put at £5000–6000 te^{-1}, pay-back period may be 8–10 years. This can give a useful picture of the sensitivity of the project to various factors, but the analysis is more valuable if displayed graphically. Thus in Fig. 3.10 the sensitivity of the cost of producing yeast biomass to the cost of raw material (glucose) and the conversion efficiency is shown. In Fig. 10.1 and 10.2 the sensitivity of costs of an immobilized enzyme process

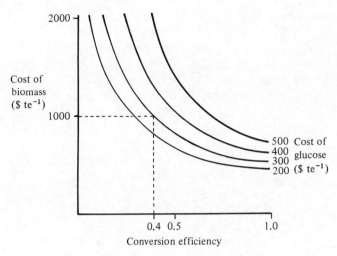

Fig. 3.10. Production of yeast biomass from glucose.

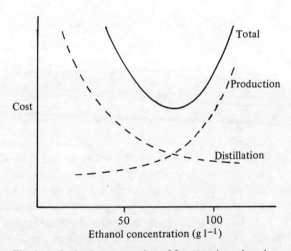

Fig. 3.11. Optimal concentration of fermentation ethanol.

(whey hydrolysis) to enzyme costs, scale and degree of hydrolysis are illustrated.

This type of analysis can also be used to determine optimum process conditions. For example in the yeast ethanol fermentation process an increased ethanol concentration in the fermenter retards ethanol production (product inhibition). On the other hand if ethanol in the broth is to be distilled, the cost of distillation varies inversely with the concentration of the incoming ethanol. A combination of these two curves results in a plot showing the ethanol concentration at which

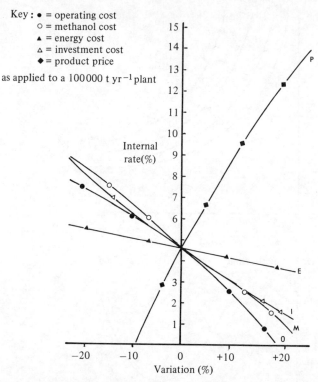

Fig. 3.12. Sensitivity analysis applied to a single cell protein process.

costs are minimal (Fig. 3.11). This concentration (80–90 g l⁻¹) is widely
employed in fermentation/distillation plants.

The data from individual plots can be combined to show in one
diagram the sensitivity of the profitability of a process to a number of
parameters, as in Fig. 3.12. This gives data from a single cell protein
project, the Norprotein process. It shows the effect of per cent
variations in the costs of a number of factors on the internal rate of
return of the process. Thus a 10% increase in methanol (raw material)
prices causes a decrease of approximately 1.5% on the internal rate of
return. An increase in product selling price of 10% increases the
internal rate of return by approximately 3.5%. The steeper the slope of
the line, the more sensitive the profitability of the process is to that
factor.

This type of analysis can provide some of the most valuable
information on the viability of a process or project proposal. It can
show the limits in any one variable which may be tolerated, it can show
where process improvements need to be made, where optimum
conditions lie and which expenditures may not in fact be needed. It is

particularly valuable in exercises such as determining the potential impact of genetic engineering, continuous culture or cell immobilization on a project or process.

Risk analysis

Sensitivity analysis may be extended to assign probabilities to particular values of the various factors being attained. For example in the Norprotein process there may be a 50% chance of the selling price reaching its predicted value, a 20% chance of it being 10% higher and a 30% chance of it being 10% lower. The probabilities can be used to calculate an expected value for the factor. Alternatively the information can be divided into optimistic, middle and pessimistic categories, or methods of dispersion such as standard deviation or the coefficient of variation may be applied.

Risk analysis is dependent upon estimates of the input figures attaining different values and the estimates may be described as subjective or objective. Objective estimates are obtained from mathematical proof or historical evidence, as for example may be the case in price fluctuations of raw materials. Subjective probabilities are merely based on intuitive judgements. New investments in biotechnology have little relevant historical precedence so much reliance must be placed on statistical normal distribution or the opinions of experts in the field.

An individual process is usually broken down into steps and each step assigned a probability of reaching a set target based on either the subjective or objective criteria. The resulting probabilities for each stage (0.2, 0.5, 0.8 etc.) may be multiplied together to give an overall probability for the whole process or fitted into a decision tree. The cumulative probability is usually based on one of the budgeting criteria described such as return on investment, net present value or internal rate of return. For example, for the fermentation project, risk factors may be ascribed to the capital cost estimate, the research costs, the variable costs and the selling price. They may be computed to give data of the type: the probability of an internal rate of return of 8% or more is 0.9, of 10% or more is 0.6, of 12% or more is 0.3 and so on. If uncertainties are too high, a project development may be shelved until further development work or a change in circumstances has reduced them.

Joint products and by-products

Most manufacturing operations in biotechnology, as elsewhere, produce two or more products. The differences may be the grade of product produced, as for example 95% or absolute ethanol, or food or industrial grade citric acid or xanthan gum. In other cases a range of different products is produced from the same raw material, as in a corn wet milling plant which may produce starch, glucose syrups, high fructose syrups and ethanol. When a group of individual products such as this is produced with each product having a significant relative sales value, the outputs are generally referred to as joint products. If the joint product has a minor value it is often called a by-product. The dividing line between a joint product and a by-product is to some degree arbitrary, but most examples in biological processes are fairly clear-cut. Cell mass from processes, such as antibiotic fermentations or yeast ethanol fermentations, is generally sold for animal feed as a by-product. The residue from distilleries using a molasses substrate (condensed molasses solubles) is also often sold as a by-product. The residues from corn processing plants, distillers' dried grains (dry milling) and gluten (wet milling) are also generally treated as by-products, although as will be seen later their value with respect to the final value of joint products is very significant.

Joint products and by-products are generally accounted for in different ways. Joint products are not identifiable as individual products until a stage in production referred to as the split-off point is reached. All costs incurred before this point (joint product costs) are allocated among members of the product group. This is a difficult exercise since often one product may not be produced without the accompaniment of another. This will be illustrated in the detailed treatment of a corn wet milling plant (Chapter 9). Costs after the split-off point can be allocated to the individual product concerned. By-products, in contrast, are accounted for by deducting their net realizable value or income from sales from the cost of the major product. In corn milling costings the sales from gluten or distillers' dried grains are generally subtracted from the substrate costs as by-product credits to give net substrate costs for, say, ethanol production. Similarly, condensed molasses solubles may be deducted from the molasses charges.

Joint product costing is a controversial area in accounting, with many different approaches adopted for inventory and income determinations. In planning it is important to note that only incremental and opportunity costs are relevant in a decision to process beyond the split-off point, such as the production of 55% fructose syrups from 42% syrups which is dependent only on the extra processing cost. If the differential revenue exceeds this cost (including appropriate capital) the additional cost is justified.

Plant location

The site for a plant is ostensibly chosen on the basis of production of the highest return on invested capital, but factors involved in the decision taking may be classified as tangible and intangible. The intangible factors can only be quantified on an arbitrary index; they include distance or time from the rest of the company, and from research and development, the community spirit and climate in the new location, and the possibility of further expansion. Some of these factors are quite significant in the development of science parks near leading universities and they account in a large measure for the proliferation of biotechnology companies in Cambridge in the UK and the Boston and San Francisco areas of the United States.

Tangible factors dominate the selection of sites for large process plants. There are several which have particular relevance for biological plants.

(*1*) *Transport.* Large plants must be sited where raw materials can be most economically transported in and products out. River or coastal sites are often favoured for delivery of commodities such as molasses by barge or ship. Plants producing high fructose syrups and ethanol are sited in corn-producing areas with good railway connections. Brazilian ethanol plants are adjacent to sugar cane plantations so that they can use bagasse for distillation as well as raw cane juice for fermentation.

(*2*) *Utilities.* Fermentation plants have a high water consumption for media preparation, product recovery and cooling. Sites must be chosen not just with year-round water supplies, but also with cool water. This factor prevented development of single cell protein plants in the Arabian Gulf where raw materials, capital and markets were available. Large volumes of effluent are also produced. Wastes with high BOD values must be treated, but water authorities may permit direct disposal of low BOD wastes under certain conditions. This factor also often favours coastal sites. Energy in the form of both electricity and steam is a very important cost input. Sites with availability of cheap energy have a great advantage. With the exception of local materials such as bagasse, energy prices are generally determined by government policies based on the availability of indigenous energy.

(*3*) *Legislation.* This may often be the most significant factor in choosing a site. In addition to national taxation policies or subsidies aimed to encourage industry, local taxes (notably property taxes) may be set to attract or discourage new industry. Environmental legislation and the availability of sites with appropriate planning permission can also be highly important.

Other factors which may have a bearing on site selection include availability of labour, proximity to markets and international taxation. Simple historical factors are often overriding, such as the company owning a site and wanting to benefit from existing facilities.

Operational analysis

Most plant operations match actual measured operating costs against predicted costs by variance analysis. Variances help feedback processes by directing attention to the areas most in need of investigation. Analysis is conducted to both improve the implementation of a decision model or to list the validity of the prediction method itself. For example a plant may have predicted a certain cost for a precipitant in a downstream process operation. If this cost is greatly exceeded it may point to faulty equipment, operator error or a factor not predicted, such as contamination by another chemical. Attention will have been focussed on the problem and remedial action facilitated.

Other factors

Many other factors are important in economic planning, decision making and plant operation. They are covered in detail in accounting and economic texts recommended for further reading. Some are also illustrated in other chapters, but it is worth making a passing reference here to some of the more significant influences.

Inflation has been an overriding factor in cost prediction in recent years, but few widely approved accounting and planning methods can deal with it effectively. In terms of cost estimation it affects most items differently and requires separate forecasts. Labour, raw materials, energy and construction costs tend to rise at different rates. Detailed indices of plant construction costs in different countries are published in chemical engineering journals. Indices of manufacturing costs are often published by different government agencies and commodity statisticians frequently make price projections with varying degrees of accuracy. An important related factor is currency exchange where fluctuations in rates can present difficulties to manufacturers operating on world-wide markets, but only producing in one or two countries. These factors may be fed into sensitivity and risk analysis.

Taxation plays a significant role in investment decisions. In particular capital and depreciation allowances can transform project viability. Different rates of taxation in different countries have been of great historical relevance in the development of biotechnology.

Different sources of capital impose different conditions upon its use, repayment, interest rate etc. The sources include equity, borrowing, internal funds, government grants and others. The most important sources in new biotechnology ventures are described in Chapter 11.

Market forecasting is essential to predict price levels, price demand curves, volumes of sales and therefore plant size, geographical locations of sales, competitors' market share and so on. Some market forecasts for new biotechnological products are included in Chapter 11. It is a very difficult area for biotechnology because many new products can now be made with interesting physical or chemical properties, but as yet no obvious uses. Earlier products from other industries, notably polymers, faced similar difficulties at one stage, yet are now often regarded as indispensable. In-depth market analysis itself is a highly sophisticated and expensive operation.

Further reading

A Handbook of Systems Analysis, 2nd edn. (1978). J. E. Bingham & G. W. P. Davies. London: Macmillan.

Cost and profitability estimation. J. B. Weaver & H. C. Bauman, in *The Chemical Engineers Handbook*, 5th edn. (1973), ed. R. H. Perry & C. H. Chilton, pp. 25.1–25.47. New York: McGraw-Hill.

Cost Accounting. A Managerial Emphasis, 3rd edn. (1972). C. T. Horngren. London: Prentice Hall International.

Capital Budgeting and Company Finance, 2nd edn. (1973). A. J. Merrett & A. Sykes. London: Longman.

Introduction to Process Economics. (1974). F. A. Holland, F. A. Watson & J. K. Wilkinson. London: John Wiley & Sons.

How to Learn from Project Disasters. True Life Stories with a Moral for Management. (1982). O. P. Kharbanda & E. A. Stallworthy. Aldershot: Gower.

Inflation in Engineering Economic Analysis. (1982). B. W. Jones. New York: John Wiley & Sons.

4 Raw materials

The significance of raw materials

The dependence of biotechnological processes upon raw material costs was illustrated by the termination of the majority of single cell protein projects based on petroleum products after the oil price rises of 1973/74. Conversely the rebirth of interest in fermentation ethanol in the United States in the late 1970s was not a consequence of advances in biotechnology or even prospects of such, but of the changes in relationship between the prices of oil and maize between 1950 and 1980 (Fig. 4.1). The technology used in ethanol production is little different from that used for centuries, with the exception now of improved distillation methods. In both the United States and Brazil surpluses of carbohydrate crops and the potential to produce more, plus political concern about balance of payments and dependence upon imported oil, have brought about large scale ethanol fermentation programmes. Just as the agrarian revolution of the late eighteenth and early nineteenth centuries is recognized as being a significant factor in the subsequent industrial revolution with all its consequences, so the improvements in

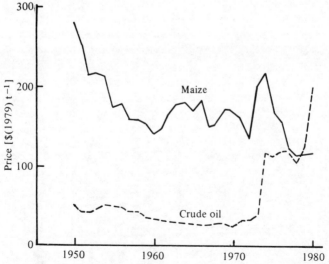

Fig. 4.1. Prices of maize and crude oil, 1950–1980. From King (1982).

74

Table 4.1. *Feedstock consumption by fermentation processes*[a]

Process	Consumption (% of total)
Ethanol	90.00
Baker's yeast	5.00
Organic acids	1.65
Amino acids	1.65
Antibiotics	0.85
Others	0.85

[a] Total feedstock consumption in 1983 was approximately 30 million te.
From Hepner & Associates (1984).

agricultural technology in recent decades with increased yields as a consequence of genetic selection, pesticides and husbandry may stimulate changes from the fossil fuel-based industry of today. As will be shown, land availability and competing uses, most importantly the production of human food, still impose severe limitations on today's technology and cereal crops.

Consumption of raw materials by fermentation processes

Products made by fermentation are dominated in volume by ethanol (Table 2.2) and production has risen rapidly in recent years. As a consequence ethanol has been estimated to consume 90% of all carbohydrate feedstocks used for fermentation, with baker's yeast accounting for 5% and all other processes the remainder (Table 4.1). The raw materials used for ethanol fermentation are predominantly maize (United States), sugar cane juice (Brazil) and molasses (Brazil, India and the rest of the world), as described in Chapter 8. Baker's yeast production uses molasses for the main part (Chapter 5). The remaining products use molasses (50%), glucose syrups (33%) and others potato and corn starches, and lactose (17%) (Hepner & Associates, 1984).

Selection of raw materials

The ability of a manufacturer to switch feedstocks is limited because of the organism's ability to use them, process optimization (i.e. yields from different feedstocks) and sometimes carry over of impurities into the products. If all these criteria are met the cheapest material is still not necessarily the best because of factors such as storage, sterilization and sometimes inhibition of the fermentation. There are examples when a more expensive substrate can bring about a more than proportional increase in revenue. For example the efficacy of *Bacillus thuringiensis* preparations is dependent upon the degree of sporulation and toxin production in the culture, which may be achieved by complex formulations of more expensive ingredients. Some impure substrates such as molasses may contain valuable growth factors such as vitamins and minerals which would otherwise have to be added. In other cases non-utilization of a significant proportion of a substrate may lead to unacceptably high BOD wastes with correspondingly high disposal costs. This is often the situation with residual lactose in hydrolysed whey preparations.

Price is nevertheless an overriding consideration, particularly at the bulk end of the market with its heavy reliance on raw material costs (Chapter 5) and product price elasticity. The production of agricultural raw materials is seasonal and therefore storage is essential. This adds to costs, as does transport. For example beet molasses may be purchased for *c.* $50 per ton at a sugar factory in North Dakota, but it costs another $20 per ton to ship it to Chicago. The water content of a material is often a significant factor in transport costs. For a raw material price to be attractive it should be low, steady and predictable. Most agricultural commodities of importance to biotechnology do not have steady or predictable prices (Figs. 4.1, 4.2, 4.4, 4.6). The impact of price fluctuations can in part be overcome by trading in futures in a commodity market, where this exists, or longer term fixed price contracts. Price is determined in part by production costs (except with wastes and by-products), by transport and storage costs and by opportunity costs (or value in alternative uses). Production costs of biomass products are associated with farm income, which may be the subject of laws or tariffs and is highly political. Opportunity costs frequently pertain to food applications and again may be in the political arena.

Availability of biomass is dictated by land use and yield, storage capability and markets. Land use is determined by price or expected price. World crop yields are increasing (Table 4.2). Storage capability favours grains while giving problems with root crops and sugar cane, which must usually be processed during the growing season. Storage can also overcome annual fluctuations in production caused by weather.

The other major criterion in substrate value is the form and availability of the carbon source. The utility of cellulosic and lignocellulosic biomass is restricted by hydrolysis to constituent sugars. Most organisms are unable to achieve this or can only do it slowly. Alternative enzyme, chemical or physical methods are too expensive. Utility of whey is restricted by the inability of most organisms to utilize lactose or galactose. Yields of products from raw materials are also

Table 4.2. *Trend in world grain production 1961–79*

Cereal crop	Mean of 1975–79 ($\times 10^9$ tons)	Change during last two decades (%)
Cereals total	1499	+34
Wheat	408	+38
Rice paddy	366	+31
Maize	352	+39
Barley	176	+43
Sorghum	62	+42
Oats	49	+2
Millet	40	+5
Rye	33	−1
Mixed cereals and others	12	+27

From: MacKey (1981). Cereal production. In *Cereals: A Renewable Resource. Theory and Practice*, ed. Y. Pomeranz & L. Munck, pp. 5–24. St Paul Minn., USA: American Society of Cereal Chemists.

dependent on the oxidation state of the substrate. The higher the degree of oxidation, the poorer the potential. Thus when oxidized substrates such as carbohydrates are compared to reduced substrates such as oil or hydrocarbons their conversion efficiencies may in fact make them more expensive even though on a weight basis they appear cheaper. This is explained in the conversions of carbohydrates and petroleum feedstocks to ethanol, acetone and butanol (Fig. 8.4). It is the reason why biotechnology, with its reliance on carbohydrate feedstocks, is predicted to have more impact on the production of oxidized bulk chemicals such as lower aldehydes, ketones and acids rather than on higher alcohols or most polymers.

The important criteria in selecting a raw material for a biotechnological process may therefore be summarized as:

(1)   Price
(2)   Availability
(3)   Composition
(4)   Form and oxidation state of carbon source

In Table 4.3 the most important potential raw materials for biotechnological processes are listed in terms of their mid-1984 price, and the price corrected relative to carbon, which compensates for (3) and (4). On this basis the most attractive candidates are corn starch, methanol, molasses and raw sugar. The corn starch price is a guessed internal price to a wet miller. As described in Chapter 9 it cannot be considered in isolation from the whole process, sales of products and by-products etc. To this price must be added hydrolysis costs for most fermentations including ethanol. It is unlikely that even

Table 4.3. *Prices of available raw materials for biotechnological processes*

Substrate	Mid-1984 US price ($ per te)	Carbon content g mol C per mol substrate	Carbon content relative to glucose (%)	Corrected price relative to glucose ($ per te)[a]
Corn starch	70–100[b]	0.44	100	64–91
Glucose	290[c]	0.4	100	290
Sucrose–raw	140[d]	0.42	105	133
Sucrose–refined	660[e]	0.42	105	629
Molasses	70	0.2[f]	50	140
Acetic acid	550	0.4	100	550
Methanol	150	0.375	94	160
Ethanol	560	0.52	130	430
Methane	n.a.	0.75	188	—
Corn oil (crude)	330	0.8	200	165
Palm oil	600	0.8	200	300
n-alkanes (n-hexadecane)	n.a.	0.87	218	—

[a] Assumes equivalent conversion efficiencies can be obtained.
[b] Approximate guessed price in a wet-milling operation.
[c] Glucose syrups on a dry weight basis.
[d] Daily spot price.
[e] US price fixed by government tariffs.
[f] On the basis of molasses being 48% by weight fermentable sugars.
n.a.: not available.

Adapted and expanded from C. Ratledge (1977). Fermentation substrates. In *Annual Reports on Fermentation Processes* 1, ed. D. Perlman & G. T. Tsao, pp. 49–71. New York: Academic Press.

when these are taken into account the effective price in ethanol fermentation in a corn wet milling operation will be anything like as high as the quoted glucose price. The raw sugar price is at an all time low. The fluctuation of this price will be considered later. It can be seen therefore why corn starch, molasses and sucrose (admittedly as cane juice in Brazil) are the most widely used substrates, particularly for bulk low price fermentation products on price relative to carbon basis. They can all be utilized by most organisms. Methanol, though cheap, can only be used by a limited range of organisms, but would, if its future availability merits it, be the target for genetic engineering of new strains producing useful products able to grow on it. The ICI single cell protein project, which was one of the few survivors after the oil price rises, is also operated on methanol. Other substrates approaching the realms of economic feasibility are the biological oils. Because of their high carbon content the cheapest oils are not much more expensive than raw sugar. In addition to corn oil, soybean oil is relatively cheap

in the US, and South American palm oil is one of the cheapest bulk
oils. Rapeseed oil is being cultivated on an increasing scale in Europe.
Biological oils are now used to a large extent where more than one
substrate is employed, for example in antibiotic and vitamin
manufacture. They will in all probability become of increasing
importance as co-substrates. All the products listed in Table 4.3 can be
stored, though the timespans vary from months (molasses) to many
years (grains for starch).

Petroleum and petroleum by-products

Interest in oil products for biotechnological processes has declined
sharply since the oil price rises, particularly with respect to cereals
(Figs. 4.1 and 4.2). Interest in enzymic conversions of petrochemicals
has also waned after poor yields and the general inability of enzymes
plus organisms to tolerate high product concentration, non-aqueous
streams. The price of crude oil has directly affected naptha and ethylene
prices (Fig. 8.2). Methane too, from both natural gas and anaerobic
digestion, will probably find greater utility in direct heating or
electricity generation. It is unattractive as a fermentation substrate
because of its poor solubility and increased fermenter costs due to the
need for flame-proofing and the avoidance of explosive mixtures with
air. It is also only utilized by a very restricted group of organisms with
complex membrane structures.

Methanol, which is principally obtained by the chemical oxidation of

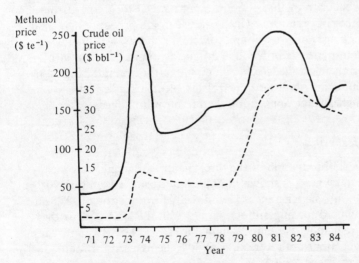

Fig. 4.2. Prices of methanol and crude oil (US), 1971–1984. Full line,
US bulk (4000 gal) methanol prices (f.o.b.) producing point as
quoted in *Chemical Marketing Reporter*. Dotted line, crude oil price.

Table 4.4. *Basis for selection of methanol as a raw material for single cell protein production by ICI*

1. High solubility in water avoids three phase transfer problems encountered with methane
2. The explosion hazard is minimal unlike methane–oxygen mixtures
3. Ready availability from a variety of hydrocarbon sources from methane to naptha
4. Ready purification with a process that avoids the carry over of polycyclic aromatic compounds
5. Less oxygen is required for metabolism and there is a lower cooling load than with methane
6. ICI manufactures large tonnages of methanol and is fully conversant with its properties and the market

From J. S. Gow, J. D. Littlehailes, S. R. L. Smith & R. B. Walter (1975). In *Single Cell Protein II*, ed. S. R. Tannenbaum and D. I. C. Wang, pp. 370–84. Cambridge, MA: MIT Press.

methane, has suffered price rises in the wake of oil price rises (Fig. 4.2) but is still extremely low priced in terms of available carbon (Table 4.3). It is interesting to look at the reasons for its selection by ICI in preference to methane as a raw material for single cell protein production (Table 4.4). This company is also investigating making the microbial storage polymer poly β-hydroxybutyrate from methanol. They have in the past been able to acquire natural gas at favourable fixed price contract rates, but this is not necessarily an advantage if they can sell methanol on the open market.

The drawbacks to methanol utilization as a feedstock, apart from the limited number of organisms which can use it, are probably that there are question marks over its future availability and uses. It may be obtained from pyrolysis of wood and liquefaction of coal, but also may be used as a chemical feedstock or fuel. It can for example be used like ethanol as an octane booster in petroleum blends. This could become a higher value application than fermentation feedstocks.

Cereal crops

The total world production of the three major world cereal crops, maize, rice and wheat, is shown in Table 4.5. It can be seen that 90% of rice is grown in Asia where it is used directly for human food. Wheat too is mostly used for human food – *c.* 70% in the USA – while the bulk of the remainder is used for animal feed. The USSR is unusual in that the proportions are reversed, with 30% being used for food and 70% for animal feed (Hill & Mustard, 1981).

From a biotechnological standpoint the most important cereal crop at present is maize. Sixty per cent of the world's production comes from

Table 4.5. *Total starch grains production* (*millions of tons*)

	All cereals	Maize	Rice	Wheat
World total	1459	349	366	386
USA	273	161	5	55
Canada	42	48	—	19
Europe	249	49	1.5	82
USSR	187	11	2.2	92
Asia	603	54	335	108
South America	64	31	13	9
Africa	66	26	8	8
Oceania	16	0.4	0.5	10

From A. J. Vlitos (1981). Natural products as feedstocks. In *The Chemical Industry*, ed.
D. H. Sharp & T. F. West, pp. 314–32. Chichester: Society of Chemical Industry
Publishers, Ellis Horwood Ltd.

the US and Canada. United States production has increased by 60%
since 1967 (Table 8.9). Approximately 30% of the output is exported,
55–60% is used for animal feed and approximately 10% is now used in
wet and dry milling. Only a tiny proportion is used directly as human
food. North America now has a surplus of maize and production is
increasing, resulting in large carry overs (Table 8.9). Approximately
70% of the weight of maize grain is starch which can be used as a
biotechnological substrate, and to produce glucose and high fructose
syrups. The details of the process are described in Chapter 9.

A successful process for wet milling wheat has not yet been
developed, despite a number of efforts. Wheat is still dry milled into
flour and wheat feed. The flour can subsequently be separated into
starch and gluten by a wet process. This two-stage scheme increases
costs and detracts considerably from the attractivity of wheat starch as
a raw material. In Sweden, which now has a surplus of over 1 million
tonnes of wheat *per annum*, a whole wheat grain to ethanol plant
(20 000 l per day) is being constructed. This operates on the continuous
Biostil process with by-product credits of CO_2, starch and animal feed.
This is however the only biotechnological plant so far to use wheat as a
raw material.

The strongest criticism of the use of cereals as biotechnological
feedstocks relates to the conflict of use of food. As shown in Fig. 4.3,
the average supply of cereal grain per world inhabitant rose by some
20% between 1960 and 1980 and the indications are that the increase is
continuing. The problem is in uneven production capacity and demand
around the world. The surplus and deficit regions have become more
and more distinct as rich nations have increased their agricultural
productivity. In addition most of the grain production of temperate

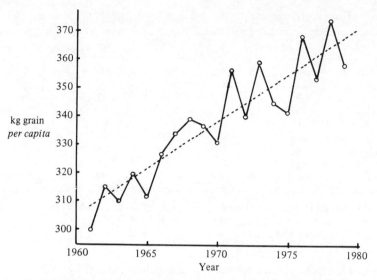

Fig. 4.3. Average supply of cereal grain in kilogram per world inhabitant over the period 1961–1979. From J. MacKey (1981). Cereal production. In *Cereals: A Renewable Resource. Theory and Practice*, pp. 5–23. St Paul, Minn, USA: American Society of Cereal Chemists.

regions is directed into animal feed which is inefficient in terms of overall human nutrition and much animal nutrition can still be satisfied by milling operations after removal of starch for biotechnology. Uses for biotechnology need not be in conflict with human nutrition, the answer to which is largely a matter of redistribution.

Sugar and molasses

World production of sucrose is in the region of 95–100 million tons *per annum*. The two sources are sugar cane grown in tropical climates, which produces approximately 60–65 million tonnes, and sugar beet grown in temperate zones, accounting for over 30 million tonnes. Much of the sugar produced is marketed in internal transactions between producer and consumer. A further substantial proportion is moved under bilateral deals between countries, such as the supplies from Cuba to the USSR or from the African, Caribbean and Pacific (ACP) countries to the EEC. Approximately 30% of the total production is traded and because of this it is disproportionately affected in price by shortages and excesses (Fig. 4.4). The world price of sugar at present is extremely low, but this is only the traded portion. A fermentation manufacturer who wishes to secure a long term contract would have to pay a higher price (Brown, 1983).

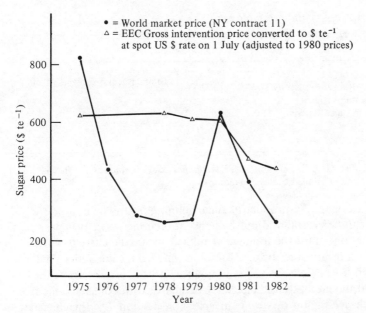

Fig. 4.4. Sugar prices, 1974–1982. Adapted from Brown (1983).

Sugar cane is bulky and deteriorates rapidly. It must be refined quickly after cutting to raw (impure brown) sugar. It can be traded in this form and is generally refined to white sugar in the country of consumption. Similarly sugar beet deteriorates rapidly and again must be refined as far as raw sugar. It is this raw sugar traded price that is quoted in Table 4.3 and as can be seen compares well with corn starch, glucose syrups and molasses at the present prices. It is rarely used as a fermentation substrate because in the past the price has been much higher. The present price is in fact considered low and most producers are running on marginal operations in order to stay in business. It must rise in the long term. The Brazilian ethanol programme runs largely on cane juice extracted directly from a cane crusher and concentrated to 20% w/v sucrose. In this form without further refining sucrose is more competitive, but can only be used in an area directly adjacent to the cane plantation and only during the harvesting period. Deterioration is too rapid to permit storage (Chapter 8).

Sugar, as other commodities, is very affected by politics. Many countries have strict import quotas and impose tariffs, hence the difference between US refined and raw sugar prices (Table 4.3) and the EEC price in Fig. 4.4. Some producers like the EEC sell off surplus refined sugar at less than its production costs on the world market. Much of the politics however is concerned with its use as a sweetener; rules may be relaxed in future in relation to fermentation uses.

Table 4.6. *Composition of molasses*

Component	% by weight	
Sucrose	25–50	⎫
Invert sugar (glucose + fructose)	1–20	⎬ Total fermentable sugars 45–50
Non-fermentable sugars	1–5	⎭
Organic non-sugar material	9–17	
Ash	10–15	
Water	15–25	

From B. P. Baker (1975). *Composition, Properties and Uses of Molasses and Related Products*. London: United Molasses Trading Co.

Molasses is the by-product of sugar refining from which no more sucrose can be crystallized out by conventional methods. It is produced at roughly one-third the tonnage of refined sugar, the current world production of molasses being 21 million tons from cane and 12 million tons from beet. Molasses varies in quality according to origin (cane or beet) and the methods and efficiency of sugar extraction. Its value from the purchaser's point of view is inversely proportional to the efficiency of extraction. As a general norm fermentable sugars have been around 48% by weight of molasses but with improved refining and extraction procedures this may be as low as 42% in some cases. Conversely inefficient refiners may sell molasses with over 50% fermentable sugars. The composition of cane and beet molasses is shown in Table 4.6. Cane molasses is generally higher in invert sugars (glucose + fructose), sodium and calcium than beet and is more palatable to animals. Both contain impurities, which may inhibit fermentations and leave high BOD wastes which may be sold as animal feed. Citric acid manufacturers use beet molasses, but must remove iron (usually by precipitation with ferrocyanide) before use. The distillation residues from ethanol plants must generally be dumped or fed to anaerobic digestors.

As a by-product molasses has no absolute cost of production and its production cannot be increased in response to increased demand; its use can only be diverted. At the moment prices are low and it does not pay molasses traders to purchase from areas such as the Sudan which have high transport costs. If prices were to rise this extra production could be traded. In Europe 30–33% of consumption is by the fermentation industry and 66% is supplied for animal feed. In the United States animal feed accounts for 78% of usage, yeast and citric acid production 16%, pharmaceuticals 4% and ethanol 2% (B. P. Baker, personal communication). Most molasses (approximately 80% of world production) is consumed within the producing country. The remaining 20% is traded across the world. This 5–6 million tons is sold largely into Western Europe, North America and Japan. It cannot

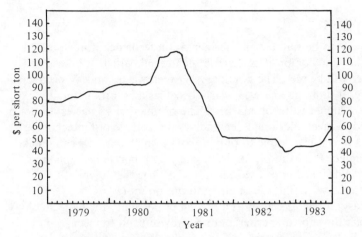

Fig. 4.5. US Blackstrap Molasses prices, 1979–1983 (f.o.b. tank car,
New Orleans). From F. O. Licht's *International Molasses Report*,
Vol. 20, No. 13, 1st September 1983, p. 125.

be stored for long periods, but is little encumbered with duties, taxes
and other restrictions. It is essentially non-political. The price is set by
the quantity available, and is based on two markets, Europe and the
United States (New Orleans). It has a value which is competitive with
other materials, notably on the animal feed market, and the price can
be extremely volatile (Fig. 4.5). It is also extremely sensitive to
transport costs.

At present cane molasses is traded for $30–35 per ton in a producing
country such as Mauritius. It is traded for approximately $70 per ton in
New Orleans, but would cost another $20 per ton farther up the
Mississippi and even more (probably over $100 per ton) in a more
remote location such as Omaha, Nebraska. The siting of a plant using
molasses as a raw material is therefore an important consideration. It
often does not pay to site a fermentation plant adjacent to a sugar
factory because of the short harvesting period and refining campaign.
For the remainder of the year the plant will be dependent upon
imported molasses and therefore is best sited at a coastal location or on
a large river such as the Mississippi where it can benefit from
reasonable transport costs year-round.

The major fermentation products obtained from molasses are citric
acid, baker's yeast, ethanol, glutamic acid and lysine. Lesser uses
include production of itaconic acid and acetone/butanol. It is not
usually used for lactic acid or for antibiotics because of the carry over
of impurities into the product and the difficulty of their removal.

Wastes

For a waste to be suitable as a fermentation feedstock it must meet the availability and composition criteria mentioned earlier. It is assumed that price will be met. The boundary between wastes and by-products is arbitrary. If utility increases, a waste becomes a by-product and acquires a higher value on markets. The production of wastes cannot be increased to meet additional demand, so the volume of biotechnology products that it is possible to obtain from available wastes can quickly be calculated. For example approximately 2 million te *per annum* of lactose in whey is dumped world-wide. This has the potential to produce nearly 1 million te of ethanol, but no more.

It is easy to regard wastes as a golden opportunity for biotechnology, but in fact the substrate criteria are usually difficult to meet. Availability of many wastes, in fact nearly all of agricultural origin, is seasonal which means that plant to handle them must either be shut down for much of the year or must be able to handle an alternative raw material. Geographical location also presents problems since collection and transport costs are often high. Many wastes are also dilute, so that water must be removed, which is expensive, or water must be transported. Dilute wastes also demand larger plant to handle them. Many wastes are not metabolizable by commercial organisms and they contain toxic components which either inhibit the process or must be removed from the product. These factors often mean that wastes can only be dealt with by sophisticated technology, the capital depreciation of which may not be justified by the added value obtained from the waste. There are in fact good reasons why wastes retain their lack of value.

All is not negative however. Carberry have constructed a plant which produces 1 million gallons of potable alcohol from whey by a *Kluyveromyces* fermentation at a large cheese plant in Balineen, Eire. Legislation on environmental pollution is tipping the balance in favour of waste treatment and production of microbial products. Several specific cases of waste utilization are covered in Chapter 10.

Future possibilities

There are numerous reports on the economic feasibility and potential of other biotechnological feedstocks whose use at present is precluded by agricultural or processing factors. Several are dealt with in specific reference to ethanol fermentations in Chapter 8, and the yields of ethanol from a variety of crops is shown in Table 8.8. Some other points merit inclusion.

The major improvements in agricultural efficiency and plant breeding

Table 4.7. *Comparison of global biomass production*

	Tonnes dry weight
Annual global biomass production	1.2×10^{11}
Utilizable wood	1.3×10^{10}
World starch production	1.1×10^{9}
World sugar production	1.2×10^{8}
World crude oil consumption	3.0×10^{9}
Lactose waste in whey	2.0×10^{6}

From U. Faust, M. Prave & M. Schlingmann (1983). An integral approach to power alcohol. *Process Biochem.*, **18 (3)**, 31–7.

have been concentrated on carbohydrate crops, notably cereals grown in temperate regions. If similar efforts were to be devoted to tropical crops it is reasonable to assume that the nutritional needs of many more nations could be met and surpluses might be available for the development of new industries. One example of a grain in this category is sweet sorghum which is now being studied in both the US and Brazil. Cassava is a tropical root crop usually produced on a small scale by subsistence farmers, frequently in mixed cropping in shifting cultivation. It is of particular potential value because if can tolerate periods of drought and poorly fertile soils and can thus be used in marginal agricultural land. In common with many other root crops the starch content is lower than grains (25–40%), but greater yield of starch per hectare can be achieved. Present yields of cassava are on average less than 10 tons per hectare, but the maximum recorded yield is 70 tons per hectare, indicating great potential as a biomass crop.

Starch need not be the only carbohydrate with raw material value. Inulin, a polymer of fructose, is obtainable in high yields from tubers of members of the *Compositae* such as the Jerusalem artichoke and chicory. It can be fermented directly by many yeasts to give good yields of ethanol, but the combined economics of agricultural and fermentation yields do not compete as yet with maize or sugar cane.

In terms of gross tonnage and potential future impact, starch is dwarfed by available wood (which is made up of cellulose, hemicellulose plus lignin, in a ratio of approximately 4:3:3) (Table 4.7). Wood amounts to 10% of global biomass production and of the remainder (comprising leaves, straw, stems etc.) the majority is cellulose and lignocellulose. It has been estimated that as much as 2.2×10^{10} te of cellulose are produced annually and that as much as 4×10^{9} te p.a. could be available for processing (Phillips & Humphrey, 1983). These polymers are conspicuously resistant to degradation (which is part of their biological function) and no economic processes for their breakdown exist at present. In form and availability they range through domestic and wood processing wastes and straw, to fast growing arboreal crops such as pines and eucalypts.

It is impossible to analyse the global and economic impacts of cellulose, hemicellulose and lignin until technical developments improve their utility. Certainly the shape of the world economy will be radically different and it can be argued that in the long term this is the most likely alternative to fossil fuels for much energy and raw material supplies. There are however many alternative scenarios, solar energy, nuclear fusion, mass pyrolysis, and even direct chemical conversions of biomass. The outcome of this competition will determine the global impact of biotechnology in the bulk product market.

Control of supply

The prices of all commodities rise and fall over any given period of time and the extent to which they do this is largely unpredictable. There are so many unknowns affecting the supply and demand for a given commodity that accurate price forecasting is impossible. Prices affecting supply include crop disasters caused by disease, floods, frosts and drought. Demand is usually influenced by overall economic conditions and the price level of a given commodity. Increased prosperity induces consumers and manufacturers to buy more goods and invest in new plant and buildings. During periods of recession the opposite happens, inventories are run down and raw materials are purchased on a much more short term basis. The opposing forces of supply and demand for a commodity are most accurately reflected in and in turn are balanced by its price.

The price fluctuations of raw materials important to biotechnological industries are shown in Figs. 4.1, 4.2, 4.4 and 4.6. Note that there are essentially two components, short term fluctuations due to limited problems of supply, crop failures etc., and a longer term trend such as is shown by the maize and oil price relationship in Fig. 4.1, and by the gradually decreasing price in real (inflation corrected) terms of biological oils through a period of high inflation (Fig. 4.6). Even given a long term trend which is favourable to a manufacturer, the short term variations can prove to be disastrous if purchases are made on the open market. Many fermentation companies who make purchases in this way are very vulnerable. The effect that these fluctuations have on a manufacturer is obviously dependent upon the total proportion of the cost of the product that is attributable to raw materials. With products such as fermentation fuel ethanol, where raw material costs are a high proportion and demand is very elastic because of the availability of ethanol from petrochemicals, these levels of price fluctuations could

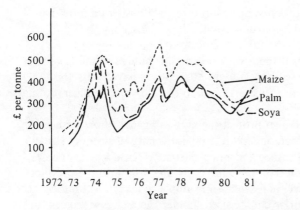

Fig. 4.6. Price fluctuations of unrefined vegetable oils, 1972–1981.
From N. Wookey & M. A. Melvin (1981). The relative economics of
wheat and maize as raw materials from starch manufacture. In
Cereals: A Renewable Resource. Theory and Practice, pp. 55–68.
St Paul, Minn., USA: American Association of Cereal Chemists.

mean that plants would be shut down because they could not even
reach their marginal costs. A similar situation would exist with
animal-feed-grade single cell protein if users could switch to an
alternative such as soybean or fish meal. At the opposite extreme lie
products such as antibiotics and other pharmaceuticals where raw
material costs constitute a much smaller percentage of the costs of the
finished goods and where demand is highly inelastic. Many products of
course lie between the two extremes. With food grade citric acid for
example, the biological route is the only one available and demand is
moderately elastic so that price rises in raw materials can be passed to
the consumer, but only to a certain extent. Demand will eventually fall,
for example in this case in soft drinks, by the use of substitutes or lower
concentrations of the product, or by reduced consumption of drinks in
response to price.

A number of options are open to enable manufacturers to reduce the
risks of price variations, but all have limitations. Manufacturers can for
example switch substrates. Most will try to develop processes which
permit this, particularly for example in changing from glucose syrups to
sucrose or molasses or *vice versa*. In some cases the prices of the two
commodities may rise and fall together because of the closeness of their
interrelationship and because manufacturers can make these kinds of
switches. Thus if the price of sugar were to rise dramatically,
production of high fructose corn syrups would rise and less glucose
syrups would be available for ethanol production. Then an ethanol
manufacturer would probably be faced with simultaneous rises in the
prices of sucrose, molasses and glucose syrups. There are also questions

of the size of the required supply and transport costs of alternative substrates to industries, particularly the large bulk commodity producers who are sited so that they can take advantage of low transport costs. Thus if there were a series of failures of the US maize crop it is difficult to see how present alternatives have either the volume or low enough transport costs to the ethanol producers to make up the shortfall.

Manufacturers can also protect themselves against short term price fluctuations in raw materials by having a diversity of product streams with differing added values and price elasticity. Their marginal costs may be covered by diverting available raw material into the products that will tolerate price increases. An example of a corn wet miller faced with a rising maize price can be used. Under these circumstances the prices of products obligatorily dependent on corn starch such as malto-dextrins, and glucose syrups in food, will probably rise more than a high fructose corn syrup where alternatives are available, and this in turn will show a lower degree of elasticity than ethanol (excluding tax allowances) where an identical substitute is available.

A third option open to a manufacturer lies in the forward purchasing of raw materials to guarantee a supply at a given price. This can be achieved by a direct fixed price contract (with a raw material supplier), or by vertical integration of the manufacturer to raw material production, or by the purchases of futures where such a market exists for the commodity concerned. Long term supply contracts have proved to be unsatisfactory. In general they have deteriorated to the point where they constitute only an initial negotiating position for both sides, because factors such as currency changes, inflation, strikes, weather etc. affect prices so much that a consumer could be competing with rivals who have bought more cheaply at the time on the spot market. Alternatively the supplier would be having to honour a contract by producing at less than cost price. Such contracts are therefore renegotiated. Vertical diversification is only of limited value because the manufacturer may be able to sell the raw material on the world market for a higher added value than he could achieve by converting it into his product.

Futures contracts are contracts which provide for the delivery of the contracted goods, but in fact only a small proportion are settled by actual delivery, that is, no physical commodity changes hands at the time of a sale or purchase. Futures markets are sometimes referred to as paper markets, but the contracts are legally binding and can be regarded as a temporary substitute for a cash transaction. Most contracts are closed out (liquidated) before they fall due for delivery by matching a sale with a purchase or a purchase with a sale of the same quantity and quality. Not all commodities have a futures market, for example molasses does not, but most major agricultural products of relevance and potential relevance here do: i.e. maize, wheat, barley, rice, sugar, fishmeal, soybeans and soybean meal.

The prime objective of a futures market is to offer protection against excessive price fluctuations to both producers and consumers of commodities. In addition to the problems of stability of supply, the price fluctuations cause great problems to the holder of stocks of raw material. The market value of a vital part of his company's assets change almost daily due to forces beyond his control, making loan securities accounting impossible and subjecting the company to the risk of takeovers. To combat this, futures contracts are hedged. This involves concomitant opposite transactions in the related futures and actuals markets. A trader who acquires a stock of soybeans hedges it by selling soybean futures and so sheds the risk of a price change he would otherwise carry.

The role of commodity market in stimulating speculation is controversial. Speculation is not dependent on the existence of these markets but they may contribute to speculative activities because goods do not have to be handled. On the other hand speculation may be regarded in some instances as being beneficial in stabilizing prices and spreading supply over good and bad periods.

Commodity markets are complex and a subject of study in their own right, but in summary their benefits are that a manufacturer can ensure a continual supply of his basic raw materials at a relatively constant price by buying the necessary futures contracts. This enables him to plan production and he can also hedge by selling futures at the same time as he buys actuals on a spot market and so cut risks of market price fluctuations. Commodities markets also give an indication of future as well as present price trends. Manufacturers must employ experts in analysing supply and demand and forecasting prices and understanding the market.

Futures markets are only of value in the short term to protect against the see-saw effect of prices (Fig. 4.1, 4.4 and 4.5). They do not affect the long term price changes which are based on production technology and utility value. This is likely to be a critical feature of biotechnology development, because the very entry of fermentation companies into these commodity markets will affect the long term price structure. Longer term planning must also take into account agricultural trends, and the planting of the correct acreage of crops possibly by co-operation with, or grants to, farmers. Much of the see-sawing of agricultural commodities is due to farmers planting a reduced acreage one year in response to a fall in price caused by reduced demand or overproduction the previous year. This then enters the political arena where governments attempt to influence economic development by selective taxation or grants. The entry of biotechnology into this arena is discussed in more detail with reference to ethanol (Chapter 8).

Summary

The future of large scale biotechnology is irrevocably associated with supplies of raw materials. Ethanol is the dominant product in terms of raw material usage at present. The oil price rises mean that biotechnology must depend on agricultural commodities or other renewable resources. At present molasses, sugar can and maize starch via wet milling are the dominant feedstocks. All feedstocks must meet criteria of price, availability and composition. Cereal crops look like being the most important short and medium term raw materials and there is evidence to suggest that this can be achieved to a certain degree without affecting human food supplies. In the long term it would appear that only cellulose and lignocelluloses can supply the scale of need for fuels if not feedstocks. In common with all agricultural commodities, biotechnological feedstocks show considerable short term price fluctuations superimposed upon long term trends. Biotechnological consumers must find ways of insulating themselves from the worst effects of these fluctuations.

Further reading

Fermentation substrates. C. Ratledge, in *Annual Reports on Fermentation Processes*, 1, (1977), 49–71. New York: Academic Press.
The Economics of Futures Trading, 2nd edn. (1978). B. A. Goss & B. S. Yarney. London: Macmillan.
Trading in Commodities, 2nd edn. (1976). Ed. C. W. S. Granger. London: Woodhead Faulkener.
Commodity prices can be obtained from newspapers such as the *Financial Times* and from *The Public Ledger's Commodity Weekly* and *Commodity Year Books*.

5 Economic analysis of fermentation processes

Introduction

Strictly speaking the present use of the term 'fermentation' is a misnomer. The word was originally used to describe anaerobic microbial processes, notably the production of ethanol from sugars. Pasteur, who first recognized the physiological role of fermentation, called it the consequence of life without air. During the growth of microbial technology to embrace aerobic processes the term has been retained although anaerobic processes are now in the minority.

Fermentation processes now cover a diversity of products in terms of applications, volumes and value, but there is a sufficient common basis to enable meaningful microeconomic comparisons to be made. The difficulty of acquiring accurate cost analysis of processes must be borne in mind here. Production costs are closely guarded secrets, particularly as a result of the oligopolostic structure of many of the product sectors of the industry. As stated earlier the reliance on estimates or projections need not be too much of a drawback since they suffice to illustrate the basic cost structure of the processes, which in detailed case studies of operating plants might be distorted by specific factors such as location, currency conversion values and local raw material or labour costs.

Fermentation has its origins in the alcoholic beverage industry and this is still apparent in terms of equipment, process design and even in terminology where for example fermenter broths may still be referred to as 'the beer'.

General cost considerations

Equipment

Fermentations are conducted in closed cylindrical vessels usually with an agitator, internal baffles and heat exchange coils, and automatic temperature, air flow, pressure, pH and foam controls. Vessels and piping are generally fabricated from stainless steel and must be engineered to fine tolerances to maintain asepsis. Bearings and valves are notable weak points. Joints in pipework must be welded and care taken in design to avoid dead spaces which might retain a contaminant during sterilization procedures. For similar reasons internal steel

93

surfaces and welds are often polished. Reactors and pipework must be leak tested, often with an inert gas.

Fermentation plant is consequently expensive, not only to purchase but also to install and commission. On the positive side it may have a working life beyond the 8–10 years generally calculated for chemical plant and usually imposed for depreciation purposes. The common design properties also mean that the equipment can be flexible and used for more than one process. Nevertheless manufacturers using fermentation processes have found that considerable expertise and man-years of experience are required to design and construct a large scale facility without crippling cost-overruns. The acquisition of this expertise forms an appreciable barrier to entry of new firms into the field. Plants are usually constructed with more than one fermenter so that the fermenter down time for repair or contamination does not affect the rest of the plant. Large plants frequently have three or four fermenters.

The capital costs of conventional fermentation plants are roughly independent of the product produced (Bartholomew & Reisman, 1979). These authors estimated that as a first approximation for a green field fermentation and finishing plant a figure of $20 to $50 per litre of installed fermentation capacity could be assumed at 1979 prices. (In 1984 prices this approximates to $30 to $70 per litre.) The value within this range depends upon the scale and the complexity of the process. These authors also estimated the scale-up of capital costs of fermentation plant to have an exponent of 0.75 compared to the 0.6 exponent common for process plants (Chapter 3).

Maintenance of asepsis

Biologists often assert that one of the great advantages of biochemical processes is that they can be conducted at low temperatures, near neutral pH and in an aqueous environment. This is in contrast to the extremes of temperature and pH and the solvents used by the chemical industry. It leads to the misconception that biological processes must therefore be less energy intensive, less hazardous and cheaper than the competition. In time this may be so, but at present the neutrality of conditions imposes a significant cost on fermentation processes in the form of maintenance of asepsis. Some fermentation processes have also suffered from bacteriophage contamination, notably acetone–butanol and the lactic acid fermentations.

In addition to the rigorous specifications on equipment, reactor vessels, raw materials and media must generally be heated to 120 °C and maintained at this temperature for 15–20 min before being cooled to operating temperature. Modern advances in heat exchange technology have reduced costs, but this expense is still considerable and has a great impact at the bulk chemical end of the market. Perhaps the greatest cost is through loss of production time or downtime through contamination. Manufacturing plants are costed on the basis of being in operation for a certain number of days per year. If this is not met the loss of production coupled with high fixed costs can easily cause the plant to run at a deficit. There is a premium on processes which can be run at high or low pH values or high temperatures so that the process organism can outgrow all potential contaminants. This is one great strength of the yeast/ethanol fermentation, which can be run between pH 3.5 and 5.0 at relatively high sugar and then high ethanol concentrations so that the yeast, if inoculated at sufficient density, can outgrow and minimize the competition without pre-sterilization.

There is also a potential hazard in the form of escape of organisms. Commercial organisms are rarely pathogenic to man under normal circumstances, but they may cause problems at high concentrations to exposed areas such as eyes or wounds. They may also be pathogenic to plants. In large scale facilities as in the laboratory they must often be treated and contained as though they were pathogens.

Aeration

Oxygen has low solubility in aqueous systems and its supply is limited by transfer from the sparger gas phase into the culture broth liquid phase, and then into the organisms. Most conventional fermentation systems use impellers to achieve oxygen transfer in addition to mixing the fermenter contents. The power requirements for impellers are high, particularly with the viscous broths encountered in fungal and polysaccharide fermentations. Power consumption is around 1.5–2.0 kW m^{-3} using disc turbines, but much is dissipated and lost as heat. The mechanical seals around the impeller shaft are also a source of contamination risk.

Conversion efficiencies

As illustrated in Chapter 3 the economics of all fermentation processes are highly sensitive to the conversion efficiencies from raw materials. In many cases strain selection and improvement work has increased these to values approaching the theoretical maximum. Costs are then dependent on the price and oxidation level of the raw material. For example the maximum conversion efficiency of a carbohydrate substrate to biomass is around 0.5 increasing to 0.8–1.0 for hydrocarbons (Pirt, 1975). Many SCP processes operate at or near the maximum recorded values. The maximum conversion efficiency of glucose to ethanol is 51% (w/w) (Chapter 8). Most fermentations can operate at around 45%.

Volumetric productivity

The fermenter volume required for a given output of product is determined by the final broth concentration of the product and the cycle time for batch processes or the dilution rate for continuous processes. The productivity P_B of a batch fermentation can be expressed as

$$P_B = \frac{V \times Y}{T}$$

where V is the volume of the fermentation, Y is the yield of the product per unit volume (final concentration), and T is the cycle time.

The productivity P_C of a continuous fermentation can in the same way be expressed as $P_C = V \times Y \times D$ where D is the dilution rate.

If for example a new plant is to be constructed to produce 1500 tonnes of product per working year of 300 days and the maximum concentration of product is 25 g l^{-1}:

(A)   *Batch process with a cycle time of 48 h*:
There are then 150 cycles per year and each must produce 10 tonnes of product.
The total capacity of reactors must be

$$\frac{10 \text{ tonnes}}{25 \text{ g}} = \frac{10\,000 \text{ kg}}{0.025 \text{ kg}} = 400\,000 \text{ litres } (400 \text{ m}^3)$$

(B)   *With a continuous process having a dilution rate of 0.05 h^{-1}*:
In a 300 day period there will be
$300 \times 24 \times 0.05 = 360$ fermenter throughputs

Each throughput must produce $\dfrac{1400}{360} = 4.167$ te product

Total capacity of reactors must be

$$\frac{4.167 \text{ te}}{25 \text{ kg}} = \frac{4167 \text{ kg}}{0.025 \text{ kg}} = 166\,680 \text{ litres or } 166.68 \text{ m}^3$$

If in the case of the batch system the cycle time can be reduced to 24 h or the product concentration increased to 50 g l^{-1} then the fermenter capacity can be halved. Similarly in the continuous system, if the dilution rate could be doubled to 0.1 h^{-1} or the product concentration increased to 50 g l^{-1} then again the capacity could be halved. The importance of these parameters to the economics of fermentation processes cannot be overemphasized. In addition, as described in Chapter 6, higher product concentrations reduce downstream processing costs.

There are often trade-offs, however. For example product concentration may only be achieved at the expense of cycle time or

Table 5.1. *Cycle time, yield and sales volume parameters*

	Bulk price ($/kg)		US sales (kg/year)		Approximate cycle time and yield (1970)	
					Time	
	1950	1970	1950	1970	(days)	g/litre
Vitamin B$_{12}$	450000	8000	20	1000	5	0.03–0.08
Pencillin G	580	23	160000	1250000	7	10–15
Monosodium glutamate	4^a	1	4300000	22000000	1.5	80–100
Citric acid	0.54	0.76	19000000	57000000	4–5	120–150

^{a} Extracted from natural products; dropped to below $2.00 per kg in 1961 with start-up of fermentation processes.
From W. H. Bartholomew & H. B. Reisman (1979). Economics of fermentation processes. In *Microbial Technology* 2nd edn, vol. II, pp. 463–96. New York: Academic Press.

dilution rate, or broth viscosity may increase with product concentration requiring extra energy inputs to maintain oxygen transfer.

The overriding importance of volumetric productivity has been a stimulus for acquiring high yielding mutants, understanding the physiology of commercial strains and improving reactor design. The impact of these parameters on the unit price and sales of four major fermentation products is shown in Table 5.1. As a consequence of low final broth concentrations vitamin B$_{12}$ remains an expensive product with low sales volumes.

Utilities

Utilities represent a major operating cost in fermentation processes. Steam is required for sterilization of medium and vessels, maintenance of sterility (steam seals) and many downstream operations such as water removal by evaporation, drying, distillation and solvent recovery. Electricity demand is high for fermenter drives, centrifuges, pumps, dryers and vacuum systems. An abundant supply of cooling water is essential, because most fermentation processes are exothermic and require cooling to a constant optimum temperature. In addition, medium and substrate streams may have to be cooled after sterilization. There is also a high demand for process water for media preparation, and dilution and washing of precipitates or crystals. The availability of cool water can have a considerable influence on overall economics and be a prime consideration in site selection. For example the development of single cell protein plants in oil-rich regions of the Middle East where waste hydrocarbons are in abundant supply has been precluded by the unavailability of adequate cooling water.

Utility charges vary from one country to another. They are lower in nations which have low priced energy, for example the United States compared to Western Europe.

Other factors

Parameters such as pH, temperature, dissolved oxygen tension, foam levels and concentrations of entry and exit gases must all be monitored and tightly controlled in many fermentations. Instrumentation costs, initial purchase, installation and maintenance costs are high. Plants often have a central control room and increasingly use microprocessor control of feedback loops. Although the cost of computers has dropped dramatically in recent years the control equipment and interfacing still present high capital costs. In fact the computer itself is now generally a minor cost in the whole set-up. As an illustration, fast response times are needed in equipment to benefit from rapid feedback times in computers. This favours mass spectrometry for gas analysis, for example. These gas analysers are more expensive than the computer and loop to control aeration or impeller speed.

Components in the growth media other than substrate may also present significant costs, particularly where the organism has specific requirements for co-factors. Foam generation is a feature of most highly aerated aqueous suspensions which contain surface active agents such as proteins, and must be controlled to reduce product loss, increase working volume and improve aeration. Both chemical agents and mechanical foam breakers are used.

Cost analysis of fermentation processes

In performing cost analyses of fermentation processes, products have been classified on a cost/volume basis starting with the cheap high volume commodities and moving through intermediate priced products such as organic acids, amino acids and polysaccharides to the higher value drugs and vitamins.

Single cell protein

The development of SCP processes based on petroleum feedstocks declined rapidly in the wake of the oil price rises of the 1970s until, in the Western world at least, only one large plant – that of ICI in Billingham, England – was ever constructed. This plant cannot be operated economically at present price levels of methanol

Table 5.2. *Operating costs of 100 000 t year⁻¹ SCP processes from methanol*

	Norprotein (%)	CTIP (%)	Hoechst/ Uhde (%)	ICI (%)	SRI Inter- national[a] (%)
Methanol	51	50	46	59	49
Other chemicals	19	28	20	17	28
Utilities	16	15	23	24	15
Labour	9 ⎫		4	—	2
Maintenance &	⎬ 7				
administration	5 ⎭		7	—	6
Methanol price	550 Nkr t⁻¹	NA	200 DM t⁻¹	NA	107 US $ t⁻¹

[a] SRI International's evaluation of a process design based principally on ICI patents.
NA = Not available.
(The calculations were made before the OPEC price rises of 1979.)
  From M. Ericcson, L. Ebbinghaus & M. Lindblom (1981). Single cell protein from methanol: economic aspects of the Norprotein process. *J. Chem. Tech. Biotechnol.*, **31**, 33–43.

feedstock and animal feed product. The company assert that the prime objective of its construction was to gain expertise in large scale fermentation technology. Many SCP plants are in operation in the Soviet Union, which has a shortage of animal feed but an apparent surplus of hydrocarbon feedstock. No data on the costings of these plants are obtainable, and neither have ICI published information on their operating costs, so all available material relates to projected plants on the basis of laboratory or pilot scale work. Nevertheless these costings do have many similarities and demonstrate the economic infeasibility of these processes under prevailing conditions.

In Table 5.2 the operating costs for five processes are compared using information gathered before the 1979 oil price rises. The processes are all for animal feed grade material, they all use methanol as a substrate for at least some of the reasons listed in Table 4.4, they are all designed for continuous operation with pressure cycle/air lift fermenters and they have initial flocculation and mechanical de-watering steps for recovering the product. They represent the later stages of SCP development and illustrate how rigorous economic pressures have caused a convergence of design. It is evident that methanol is the most important operating cost and this is amplified when the production costs for these processes are re-calculated using the post-1979 methanol prices of $150 t⁻¹, or higher.

The products of these processes have higher protein contents than soy bean meal and are rich in vitamins, but will not generally attain the price levels of fish meal. Soy meal sells from around $200 per ton and fish meal at $350–400 on the basis of protein content and quality. Even at price levels approaching fish meal with methanol prices at $150 per ton and higher, these plants will barely cover their operating costs. Even during the 1960s, at a time when methanol prices were much lower in relation to protein feeds, the economics of such plants were so dependent on substrate that maximum conversion efficiencies were essential, and all other costs, i.e. utilities for aeration, de-watering and drying, had to be minimized.

When capital costs are considered the schemes become entirely academic. These costs relate to plants with 100 000 te p.a. capacity which in later 1970s prices would have cost $60 000 000–100 000 000 and today would probably be in excess of $150 000 000. The ICI Pruteen plant was designed to produce 70 000 te p.a. and is estimated to have cost in the region of $100 million (1980 prices), but it was a pioneer. Subsequent plants would be predicted to be cheaper. Even so repayment of capital costs and interest even on a $100 million plant would be in excess of $15 million p.a. or $150 per te of product to satisfy most companies' requirements for return on investment. Clearly, at present methanol prices, the SCP product will have to sell for a price well in excess of $500 per te for such plants to be considered.

A breakdown of capital costs for the ICI Pruteen process (Table 5.3) shows that the downstream operations of de-watering and drying make a larger contribution than the fermenter and compression equipment for achieving gas transfer. When storage, packaging, services and effluent treatment are included there is a spread of costs over the whole operation with no one component reaching 20% of the total. This is interesting in view of the publicity and engineering achievements associated with the design, construction and erection of the 150 m^3 fermenter. In the Norprotein process, which is similar in concept to the Pruteen process, comprising one large airlift fermenter with a capacity *c.* 2000 m^3, a capacity scale exponent of 0.75 has been estimated in agreement with Bartholomew & Reisman (1979); however it must be remembered that the maximum size of most of the equipment used is already reached at this scale. If four fermenters are installed to achieve the same total capacity of 100 000 te per year, investment costs increase by 9% (Ericsson, Ebbinhaus & Lindblom, 1981).

Single cell protein processes are of economic interest because they have pushed today's fermentation technology to its limits for a cheap bulk product. The stringent cost targets have developed airlift fermenter technology and forced economies of scale and economies in downstream processing. Plants of this size are also operating and are under construction in Eastern bloc countries. For example, a large plant at Mozyr in Byelorussia with a designed capacity of 300 000 te

Table 5.3. *Capital costs of the Pruteen process (1980)*

Item	% of total cost
De-watering	19
Off-site services	16
Fermentation	14
Drying	12
Storage and packing	12
On-site services	11
Compression	9
Effluent treatment	4
Raw materials	3

Data from ICI as quoted in D. Fishlock (1982).
The Business of Biotechnology, p. 104. London:
Financial Times Business Information Ltd.

p.a. is being brought onstream. It is a joint development between
several countries. Another 60000 te p.a. plant growing a *Candida* yeast
on paraffins is operating in Curtea de Arges, Romania. It has four
airlift fermenters each with 1260 m³ capacity.

The future of SCP in the West is uncertain. The cost breakdowns of
processes may be approximated to 30% capital, 40% substrate and 30%
others (media, sterilization, downstream operations). It is very difficult
to see where further significant economies in capital and operating
costs can be made. Substrate costs may be cut by new materials (cheap
degradation of cellulose and lignocellulose being the prime candidate)
or by utilization of wastes. It is here that perhaps the best short term
applications may be found. The economics of a process using
confectionery wastes is described in Chapter 10. The other opportunity
is to look at more expensive end uses, e.g. human food, which can carry
the costs. These tended to fall from favour because of regulatory
problems concerning hydrocarbon residues carried through into the
product and the bad publicity from the BP Toprina plant in Italy (see
Fishlock, 1982). There were also problems in developing products with
the taste and textural requirements for human consumption. In Britain,
however, Rank Hovis McDougall pursued their *Fusarium*
'Mycoprotein' product which uses a glucose substrate. They obtained
regulatory approval and succeeded in producing great improvements in
textural and flavouring characteristics. The project was suspended for
some time, because having spent an estimated £30 million, much of it in
obtaining regulatory approval, at least an equivalent amount was
required for the construction of a manufacturing plant and the market

future of the product was uncertain. This company have since announced a joint venture with ICI to develop this project with the latter's fermentation technology, and the product has had a favourable public response, so the Billingham plant may yet produce a food grade product.

Ethanol

Ethanol production for potable use is the oldest and still the largest fermentation industry. For industrial and fuel uses fermentation has been replaced by the oxidation of ethylene obtained from petroleum naphtha in most parts of the world. Fermentation survived in some nations, notably India, when supported by government programmes and is now undergoing a revival, again only with government support, as a result of oil price rises since 1973. The details of ethanol production in terms of energy balances, contribution to world energy supplies, competition with petrochemicals, political and other aspects are discussed with other energy issues in Chapter 8. For the purpose of this chapter the aim is to compare costs of the ethanol fermentation in relation to fermentation costs of other products.

Industrial/fuel ethanol production is now greatest in the United States and Brazil and is growing rapidly in both countries. It is therefore appropriate to examine production costs in these two countries. All the processes involve *Saccharomyces* yeast at present. In Brazil the predominant raw material is the juice extracted by crushing sugar cane; it is used directly without further treatment. Molasses is used on a smaller scale, but cassava is recognized as being a raw material with great potential (Chapter 8). The economics of Brazilian ethanol plants using cane juice and cassava are compared in Table 5.4. The cassava costs are taken from small scale work but are extrapolated to the scale of the cane process. Cassava can be stored, but cane must be used directly after harvesting and is therefore seasonal. The capital costs (depreciation) of the cassava plant are lower because the same output can be achieved over 330 days per year rather than the 180 permissible with sugar cane. The substrate in cassava is starch which must be extracted and hydrolysed to glucose or maltose before fermentation. In the case of sugar cane the plant residue or bagasse from the crushers is burnt to provide energy for distillation. This results in a higher enzyme, chemicals and utilities charge for cassava. Otherwise the cost breakdown of the two processes is similar. Raw material costs are the largest contribution at around 60% of the total in both cases.

In the United States corn (maize) is the favoured substrate, but it is produced in three distinct categories of plant: small farm units, and large scale dry and wet milling operations. The economics of farm operation and a wet-milling plant are compared in Table 5.5. The farm

Table 5.4. *Economics of production of anhydrous ethanol from cassava and sugar cane in Brazil using plants of the same daily capacity*

	Cassava distillery (hypothetical)		Sugar cane distillery	
	$/m³	% of total cost	$/m³	% of total cost
Cassava roots at $33.3 ton⁻¹	228	60.1	—	—
Sugar cane at $13.6 ton⁻¹	—	—	204	57.0
Enzymes, chemical & utilities	60	15.8	6	1.7
By-products (credit)	(16)	(4.2)	(15)	(4.2)
Labour	13	3.4	15	4.2
Maintenance, materials, supplies, insurance & admin. expenses	18	4.8	25	7.0
Taxes	15	4.0	24	6.7
Depreciation	32	8.4	49	13.6
Net operating profit	29	7.7	50	14.0

Basis: 150 m³ per day cassava distillery operating 330 days p.a. (49 500 m³ p.a.)
150 m³ per day sugar cane distillery operating 180 days p.a. (27 000 m³ p.a.)
From: V. Yang & S. C. Trindale (1979). *The Brazilian Gasohol Programme Development Digest*, **17** (3), 12–24.

costs are optimistic and may be questioned, but more on a dollars per gallon basis than on percentage breakdown. As with the Brazilian examples raw materials costs predominate. In corn wet milling they are reduced by by-product utilities; these are discussed in detail in Chapter 9 and are put by some authorities as lower than the 50% cited here, while by others as high as 63% (Keim, 1983). Nevertheless, raw materials are still the largest charge.

As with single cell protein, improvements in ethanol recovery, notably distillation, plant design and utility costs, can all benefit process economics, but they are secondary to raw material considerations.

It is interesting to compare these costs with those for a potable alcohol fermentation (Table 5.6). These relate to data collected from British Breweries in the mid-1960s. The excise duty figure is very high: in Britain it is now approximately 30% of the retail price and in most other Western countries in a 10–20% range, though often local taxes may be added. Even so the cost structure is very different from the fuel alcohol cases. Raw materials are much less significant, even though in absolute terms they are more expensive, comprising as they do malted barley, hops and refined sugars. Distribution costs are much higher because of smaller volumes and many more diverse outlets. Because the costs refer to beer there is no distillation step, but it is reasonable to assume that with spirits the cost breakdown would show a greater

Table 5.5. *Ethanol production economics in the United States* ($ *per gallon*)

Item	Small farm scale[a] 150000 gal p.a. Selling price $1.74/gal		Large wet-milling plant.[b] Corn at $3.00 per bushel. Selling price $1.70/gal	
	$ per gal	% total cost	$ per gal	% total cost
Corn (gross)	0.92		1.20	
By-products	—		0.60	
Net corn cost	0.92	61	0.60	43
Enzymes + yeast	0.04	3	0.10	7
Labour	0.14	9	0.05	4
Electricity	0.02	1	—	
Straw	0.07	5	—	
Total energy	—		0.25	18
Miscellaneous	0.12	8	0.10	7
Total variable expenses	1.31	87	1.10	79
Depreciation	0.05	3	—	
Interest	0.15	10	—	
Total capital costs	0.20	13	0.30	21
Total production cost	1.51		1.40	
Pre-tax profit	0.23		0.30	

[a] *From*: *Fuel from Farms*: *A Guide to Small Scale Ethanol Production*. (1980). Solar Energy Research Institute, US Dept. of Energy, Oak Ridge, Tennessee.
[b] *From*: Drexel Burnham & Lambert Inc. *Nutritive Sweetener Industry Outlook*, Jan. 6th 1983.

resemblance to beer than to industrial ethanol, particularly with respect to excise duty.

The selling price of alcoholic beverages bears less relationship to production price than is the case for industrial ethanol. Demand is far less elastic and is dependent on location, organoleptic properties, fashion and so on. Potable alcohol retails for prices in the region of $10–20 per litre (supermarket prices) as opposed to $0.45–1.0 for 95% or absolute industrial ethanol. In short it is a different business. A brief comparison of the two markets is outlined in Table 5.7. Many of the traditional food fermentation industries, such as yoghurt, cheese and soy sauce, resemble potable alcohol in these respects.

Table 5.6. *Structure of brewing costs*

Item	% of total cost
Brewery materials	8.6
Direct labour	7.2
Depreciation	2.8
Other costs (purchasing cleaning materials, power etc.)	12.9
Total production costs	31.5
Excise duty	61.4
Distribution costs	7.1

Based on National Board for Prices and Incomes (1966 report).
Data from average costs for 40 UK breweries in 1965.
From: C. F. Patten (1971). Brewing. In *Economics of Scale in Manufacturing Industry*, p. 74. Cambridge University Press.

Table 5.7. *Comparison of beverage and fuel ethanol*

Beverage	Fuel
Good profit margins	Only profitable with political intervention (Tax credits etc.)
Flavour constraints	Flavour not relevant
No dehydration necessary	Dehydration usually necessary
Energy consumed in production of some importance	Energy paramount

From: Coote (1983).

Yeast biomass

Yeast is produced in large tonnages world-wide (Table 5.8) for food and feed (dried yeast) and for baking. It differs from the single cell protein examples described earlier in that most processes are on a smaller scale and use conventional technology. Plant is often old and investment amortized. Feed yeasts (*Torulopsis* and *Candida* strains) are usually grown on molasses or spent sulphite liquor (Chapter 10), but occasionally wood hydrolysates are used. Yeasts recovered as by-products of alcohol fermentations both in grain and molasses distillation and in brewing are also sold for animal feed, often after de-bittering, and are included in the totals in Table 5.13. Baker's yeast (*Saccharomyces cerevisiae*) is more of a premium product. It was formerly grown on grain and potatoes, but today molasses is mostly used. A variety of processes are used in both feed and baker's yeast production; some are purely aerobic while others have an anaerobic or

Table 5.8. *Estimated annual yeast production, 1977 (dry tons)*

Location	Baker's yeast	Dried yeast[a]
Europe	74000	160000
North America	73000	53000
The Orient	15000	25000
United Kingdom	15500	[b]
South America	7500	2000
Africa	2700	2500

[a] Dried yeast includes food and fodder yeasts.
[b] None reported.
Production figures for USSR not reported.
From: H. J. Peppler (1979). Production of yeasts and yeast products. In *Microbial Technology* 2nd edn, Vol. 1, p. 159. New York: Academic Press.

Table 5.9. *Costs of yeast biomass production*

Component	Requirement per te feed yeast (92% dry matter)	Unit cost	Cost per te yeast ($)	Percentage of selling price ($1000/te)
Variable costs				
Molasses	4000 kg	$70/te	280	28
Water (20 °C)	500 m³	2c/m³	10	1
Steam	10 te	$20/te	200	20
Electricity	1200 kWh	5¢/kWh	60	60
Sulphuric acid	40 kg	$0.09 /kg	3.6	0.36
Diammonium phosphate	80 kg	$250/te	20	2.0
Urea	150 kg	$220/t	·33	3.3
Magnesium sulphate	5 kg	$0.30 /kg	1.5	0.15
Antifoam oil (crude lanolin)	10 kg	$1.20/kg	12	1.2
Total variable cost			620	62
Fixed costs (plant overheads, labour)			100	
Interest and depreciation at 20% of capital cost p.a.			100	
Total manufacturing cost			820	

Plant capacity 6000 te yeast p.a.
Capital cost $3.0 million.
Data adapted from Paturau (1982).

reduced aeration stage producing ethanol as a by-product which is recovered from broths by distillation after cell removal. Some processes are continuous, others are batch. A high proportion are now fed batch with extended addition of substrate and removal of cells. There is also considerable variation in inoculum size; often small seed fermenters are used to reduce growth time in the main vessel to around 12 h.

The economics of feed yeast production are very sensitive to substrate price (Table 5.9). In many sugar cane growing regions molasses is sold at present for \$30 te^{-1}, but it is fairly expensive to transport and distribute, resulting in prices of around \$70–80 te^{-1} delivered in importing countries. This difference adds \$200 te^{-1} to the variable cost of yeast biomass production. Sulphite liquors are much more attractive where available. Utility charges are also high, partly through the need to aerate the culture, but mostly through sterilization of media, in cell separation and in drying the product. Other medium components are relatively inexpensive. The fixed costs such as labour and plant overheads have been guessed at \$100 te^{-1}. The total costs are much higher than for SCP because of the smaller scale, batch operation, less efficient drying and more expensive substrate. Even with these high costs feed yeast production can be economically viable (though probably not in a new plant) because the product can retail for around \$1000 te^{-1} (dry solids) in Europe and North America.

Baker's yeast is frequently sold as a cream containing 27–29% dry matter which is pressed into pieces of different shapes and weights for bakery and retail sales, though it can also be dehydrated to active dry baker's yeast. Coloured impurities from the broth are removed by washing and centrifugation. The fermentation economics are approximately similar to the feed yeast case with lower utility costs for cream production, but higher packing and distribution costs which will comprise a high proportion (probably at least 25%) of the selling price.

With such a variety of processes, many retained for traditional reasons, and great variation in substrate costs it is impossible to define production costs any more accurately. Baker's yeast production resembles brewing in that there is a spread of costs through factors such as distribution to many small customers and emphasis on the traditional flavour constraints associated with a foodstuff.

Carboxylic acids

Fuel ethanol and single cell protein represent the cheapest products obtained by fermentation processes. In terms of price the next group of products are the carboxylic acids, particularly citrate, gluconate and lactate (Table 5.13). All are long established products originating before the Second World War. The existing markets are dominated by a small number of important and efficient producers who have kept their practical know-how fairly secret. For example the majority of the US domestic consumption of citric acid is supplied by Pfizer Inc. and Miles Inc.

World production of citric acid is around 300000 te p.a., but installed capacity has been put as high as 500000 te p.a. In the West the chief producers of citric acid are located in the US, Italy, Belgium and France, but many Eastern bloc countries are now substantial producers. Despite the present overcapacity new citric acid plants may be constructed in developing nations where the raw material (molasses) is cheap, and demand – particularly in soft drinks and foods – is rising.

Citric acid is generally produced by *Aspergillus niger* using beet molasses as a substrate. The molasses is usually treated with ferrocyanide to remove Fe^{2+} ions which inhibit the fermentation. Two methods, surface culture (in trays or shallow tanks) and submerged aerated culture, are used. The economics of the two processes have been compared in detail (Schierholt, 1977; Sodeck et al., 1981). In the older surface process the fermentation is carried out in large trays arranged in several layers one above the other and side by side. Fermentation takes place on the surface of the solution in a mycelium layer, and the depth of the solution must be kept shallow. This requires a large floor space and hence higher investment costs in buildings. A mechanical system operates filling and discharge of the trays, but even so labour costs are higher than in the submerged process which uses large fermentation tanks with agitation and aeration. The mycelial culture becomes viscous and high energy inputs are required to achieve oxygen transfer. The equipment costs are higher in the submerged system; raw material costs are roughly similar in both methods. High conversion efficiencies of up to 85% yields from starting sugars have been reported (Atkinson & Mavituna, 1983), but probably 70–80% is more usual. The fermentation takes 4–5 days. The surface process may have some economic advantages in small plants of less than 1000–1500 te p.a. but in general it is being displaced by submerged fermentation in larger plants.

An approximate breakdown of the costs of a submerged citric acid fermentation is given in Table 5.10. These figures have been derived by applying 1984 prices to the materials balance of Paturau (1982), who concludes that it is unlikely that a competitive new plant could be built with a capacity of less than 2500 te p.a., and that a cheap source of sulphuric acid is an important advantage. He puts the cost of such a plant at about $3.5 million. Many older plants in operation will not have the same capital depreciation and interest charges. Citric acid is collected from broths by precipitation with lime and subsequent treatment of the calcium citrate with acid.

It can be seen that costs are spread much more evenly over raw materials, downstream recovery operations, utilities and capital charges

Table 5.10. *Costs of citric acid production*

Component	Requirement per te citric acid	Unit cost	Cost per te citric acid ($)	Percentage of selling price ($1560/te)
Variable costs				
Molasses	3000 kg	$80/te	240.00	15.3
Sulphuric acid	800 kg	$90/te	72.00	4.6
Active Carbon	25 kg	$500/te	12.50	—
Electricity	3800 kWh	5 c/kWh	190.00	12.1
Nutrients	5–15 kg	$200/te	2.00	—
Lime	450 kg	$40/te	18.00	—
Water	25 m³	2 c/m³	0.50	—
Steam	19 te	$20/te	380.00	24.3
Total variable cost			915.00	58.6
Interest and depreciation at 20% of capital cost p.a.			280.00	18.0

Plant capacity 2500 te citric acid p.a.
Capital cost $3.5 million.
Note: Labour costs are not included.
Data adapted from Paturau (1982).

than is the case with ethanol or single cell protein. Similar overall cost
breakdowns apply to lactic, gluconic, and itaconic acids. All can be
obtained in high yields from carbohydrate starting materials, though
more expensive glucose syrups replace molasses. In the case of lactic
and itaconic acids this is because of the difficulty of removing colour
from the product. Extraction and purification are multi-step processes,
and costs are high in terms of chemicals and utilities. Capital costs of
plant come high partly because of the need for corrosion resistant
materials such as high grade stainless steel. (With lactic acid wooden
fermenters were used at one time.) Production costs are also dependent
on the grade of material supplied. Lactic acid for example is supplied in
technical, edible, plastic and USP (pharmaceutical) grades with
decreasing impurity levels and increasing costs. Fermentation lactic acid
has been replaced by the racemic product obtained from
acetaldehyde + HCN via lactonitrile in many low grade industrial
applications. Overall only about 50% of present world lactic acid
production of 40000 te p.a. is now met by fermentation. As with citric
acid there is world-wide fermentation overcapacity and a static if not
slightly declining market.

The unfavourable market conditions and low profit margins in
organic acids have led to little investment in research and development
of new equipment, processes or strains. This enforced conservatism
extends to vinegar manufacture which uses *Acetobacter* species.

Amino acids

In contrast to organic acids the production of amino acids by fermentation has been a growth industry. Monosodium glutamate production, formerly achieved by hydrolysis of protein such as wheat gluten, has grown in volume and dropped in price since the introduction of the fermentation route (Table 5.1). Similarly the growth in L-lysine production has been dramatic since the development of a successful fermentation. As described in Chapter 2 amino acids display classic price/volume relationships. Where large markets such as flavour enhancement or supply of an essential amino acid for animal feed exist, prices have dropped and production volumes have risen. Where applications are limited to therapeutic or laboratory chemicals, volumes have remained low and prices high. The exception is L-tryptophan where a potential feedstuff market has not yet been met by reduced production costs. Genetically engineered strains may change this. Similarly phenylalanine now has a large volume application in the production of Aspartame. Supply seems to be rising and prices falling due to the application of genetic engineering.

Monosodium glutamate and lysine are the most significant products in terms of sales volume. Yields of MSG have been improved by the isolation of *Corynebacterium glutamicum* strains with mutations in the citric acid cycle and with phospholipid-deficient 'leaky' cytoplasmic membranes when grown in the absence of biotin. Yields of lysine have been increased by overcoming feedback repression of biosynthesis through developing resistance to non-metabolizable analogues of the amino acid. The strains are examples of the successful application of what can now be described as conventional genetics. As a result yields of 40% (w/w) MSG from fermentable sugars in molasses and over 30% lysine from glucose are now obtained. It is reported that a new lysine plant at Cape Girardeau Missouri (Kyowa Hakko) will use molasses as feedstock, but most manufacturers to date have used glucose syrups because of the difficulty in removing molasses impurities from the product. These yields reduce substance costs to approximately 15% of the selling price. Separation, purification and crystallization of the products from the broth is still a multi-stage and relatively expensive process despite increased concentrations in the 40 g l⁻¹ range, giving an overall spread of costs amongst utilities and reagents similar to citric acid. The cost of the new Kyowa plant is put at $45 million for a capacity of 7500 te p.a. This would put depreciation and interest charges at *c*. $1000 per te on a 10-year write-off basis.

Amino acid production is concentrated among a few suppliers, notably Ajinomoto (which is responsible for approximately half the world MSG supply), Takeda and Kyowa. The process technology and capital investment required for a new plant with sufficient economy of scale are barriers to the entry of new firms into the market.

Microbial polysaccharides

Microbial polysaccharides are representative of a group of products often referred to as speciality chemicals with prices and volumes between the cheap bulk chemicals and the high value, low volume specialist products such as antibiotics and vitamins. Only two microbial polysaccharides to date – xanthan and dextran – have significant commercial production.

Xanthan is the extracellular polysaccharide produced by *Xanthomonas campestris*. Its most important property is its ability to control the rheological properties of fluids and it is used as an emulsifier, thickener, stabilizer and as a component of oil drilling muds. It has also been reported to have potential in enhanced oil recovery. The present world-wide market is for approximately 10000 te xanthan with a sales value in the region of $100 million, but there are considerable differences in prices, purity and production costs between the industrial grade products, particularly between drilling muds and the food grade products. New production capacity is coming onstream with for example a plant in France to produce 3000 te p.a. (Rhone Poulenc).

Unfortunately, it has proved impossible to obtain even an approximate cost schedule of xanthan production for publication. There is, though, a sizeable patent and research literature on it from which some estimates may be made

Xanthomonas campestris is generally grown under nitrogen or possibly sulphur limitation in batch culture, in a stirred tank reactor, using glucose and protein, protein hydrolysate or sometimes yeast extract as substrates. Growth and xanthan production occur simultaneously, resulting in a highly viscous and pseudoplastic broth. The final polysaccharide concentration is usually 25–30 g l^{-1}. Higher concentrations require increased power inputs to impellers to achieve oxygen transfer in the broth. Product recovery costs are higher at lower concentrations. Xanthan is precipitated from broths using isopropanol, and cells are subsequently removed by heat treatment. Isopropanol is recovered by distillation and re-used, but there are inevitably some losses, and it still represents a significant cost. Drying the product is also expensive and food grades must be purified free of isopropanol, metal ions and other contaminants. The production plant, too, is expensive. It could cost in the region of $15–20 million for a 1500 te p.a. facility, although economies of scale might be achieved at greater capacities. Industrial grade xanthan is sold for $9000 te^{-1} (f.o.b.), food grades sell for up to $14000 te^{-1}.

In practice, conversion efficiencies of 60–70% can be obtained from a glucose substrate and, even though this is supplemented with protein, total substrate costs probably only amount to 5–10% of the selling price, dependent upon the price of glucose or glucose syrups in the country concerned. Utility charges are high, perhaps 20% or more of the selling price, because of the aeration problems and the energy intensive recovery process. Capital charges, particularly on a new plant, will also be a very significant component of total costs, perhaps up to 30% of the selling price, if the plant is written off over 10 years or less. When other expenses, such as packaging, effluent treatment and chemicals employed in recovery are included, it can be seen that the trend towards a greater spread of costs in more expensive fermentation products is continued, with greater emphasis on downstream processing.

There is some scope for reducing xanthan costs, but it is probably fairly limited. Two stage processes, in which growth and xanthan biosynthesis are separated, can reduce the oxygen supply costs of the fermentation, but they will involve increased capital expenditure. The organism can be grown in continuous culture, but only at slow dilution rates (c. 0.05 h^{-1}), so reductions in capital costs here are fairly marginal and must be balanced against down time caused by contamination. Contamination risks are high in a medium of this composition and can only be overcome by very well-designed plant.

Dextran is an extracellular 1–6-linked polysaccharide of D-glucose produced by *Leuconostoc* strains growing on a sucrose substrate. It is also often manufactured enzymically by dextransucrase isolated from these strains. It can be used as a stabilizer for ice cream, sugar syrup and other confectionery but is generally too expensive. Its only important use is as a blood plasma volume expander. Dextran is recovered from fermenter broths by precipitation with ethanol or methanol and is purified to clinical standards. It is also separated to desired molecular weight ranges. The total world-wide production is in the 100–200 tonne range. Plants are therefore tiny, probably no more than 30 te p.a., with the purification stage accounting for a very high proportion of costs.

Penicillin

Penicillin is now produced by high yielding mutants of *Penicillium chrysogenum* in fed-batch fermentations. The spectacular rise in productivity of penicillin manufacture and the corresponding fall in price has been one of the great success stories of fermentation technology (Fig. 5.1). The final broth concentration in penicillin

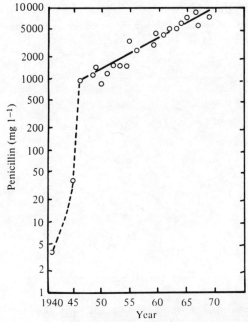

Fig. 5.1. Changes in penicillin broth potencies with time. From
A. L. Demain (1971). Overproduction of microbial metabolites and
enzymes due to alteration of regulation. In *Adv. in Biochemical
Engineering 1*, p. 113, ed. T. K. Ghose *et al.*, p. 113. Berlin: Springer-
Verlag.

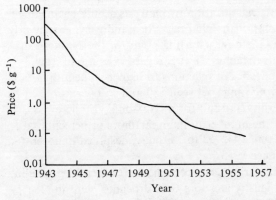

Fig. 5.2. History of the cost of penicillin. From P. P. King (1982).

fermentations has risen from approximately 2 units ml^{-1} to 50000 ml^{-1}.
This is an increase from 1.2 mg ml^{-1} to 30 g l^{-1} of penicillin G (sodium
salt), and it has been achieved by strain selection, understanding the
physiology of the process, medium composition and reactor design.
Changes in penicillin broth potencies over time are shown in Fig. 5.2.

Table 5.11. *Manufacturing cost of benzyl penicillin*

	Per cent of total cost		
	Fermentation	Recovery	Total
Variable costs:			
Raw materials			
Carbohydrate phenyl-acetate etc.	44	—	
Solvents	—	5	49
Utilities			
Steam electricity	8	4	12
Fixed costs:			
Labour, maintenance overheads, depreciation	26	13	39
Total	78	22	100

Costs refer to 100 m³ fermentation.
From: J. de Flines (1980). *Biotechnology: a Hidden Past, a Shining Future*, pp. 12–17. The Hague: TNO.

Penicillin was originally produced in trays or glass bottles, but is now obtained from 100 000 l fermenters.

The fall in price of penicillin and the increase in volume of production have led to its classification by many as a bulk chemical. This is largely a standpoint of the pharmaceutical industry, particularly because much penicillin G is now a starting point for the synthesis of the substituted penicillins (Chapter 7). It is regarded as a bulk raw material for an expensive series of reactions to higher value antibiotics. The bulk prices of the drug range between $30 and $50 per kg depending on grade, sterility and salt form (Na^+ or K^+). There has been a concentration of the number of manufacturers of penicillin, it having declined from 100 in 1945–53 to 35 today, despite a hundred-fold increase in production since 1953.

A summary of manufacturing costs of penicillin G (benzyl penicillin) is shown in Table 5.11. This relates to a Gist Brocades plant in Delft, The Netherlands, using 100 m³ fermentations with 200 h fermentation time (fed batch) and 15 h recovery of penicillin from the fermenter. Potassium phenylacetate is added as a precursor for the side chain of benzyl penicillin. Fourteen fermenters are used to maintain continuous recovery operations. The recovery process comprises separation of mycelia from broth through a continuous rotating filter and extraction of the broth plus washings with butylacetate in a countercurrent extractor. Penicillin is precipitated from butylacetate as the potassium salt, filtered, washed with butanol, filtered and dried giving a product of 99.5% purity which sold (in 1980) for about $35 kg^{-1}.

This recovery process is efficient and cheap, giving only a 22% contribution of manufacturing costs. The biggest single item in costs are the substrates for fermentation (44%), though it must be noted here that some phenylacetate can be recovered if the next step for penicillin is de-acylation to 6-aminopenicillanic acid for synthesis of substituted derivatives (Table 7.9). Similarly the solvents used in recovery are recycled by distillation. Variable costs make up 60% of the total and fixed costs 40%, though the latter would be higher for a new plant, as described by Cooney (1979). These costs do not include effluent treatment of the high BOD wastes, particularly the spent fermenter broth, which could be an appreciable addition.

High value products

Nearly all other antibiotics in clinical use are more expensive than penicillin many by orders of magnitude. Some, for example the cephalosporins, have prices in the thousands of dollars per kg range. The cephalosporins and semi-synthetic penicillins are expensive because the starting antibiotic is produced by fermentation, but is then modified by a series of chemical or enzymic steps. These groups of antibiotics have the fastest growing share of the market. Other antibiotics, although manufactured by fermentation in the same way as penicillin, are more expensive because conversion efficiencies from substrates are low, final broth concentrations are low (often less than 1 g l^{-1}) and downstream processing is more complex and more expensive.

A similar situation exists in vitamin production by fermentation. It has so far proved impossible to increase the conversion efficiencies, yields or downstream operations to achieve the prices and volumes of other fermentation products. In some cases this failure has led to a chemical route supplanting the biological one. For example riboflavin (vitamin B$_2$) was formerly made exclusively by fermentation, but now a chemical synthesis from ribose is used for most pharmaceutical grade material. The fermentation product from the mould *Ashbya gossypiae* is still used for animal feed supplements where it is sold in dried whole cells without purification. Only one manufacturer in the United States (Merck) still uses the fermentation route. This product is sold for $50-55 per kg on a riboflavin basis for animal feed. It accounts for one-fifth to one-sixth of current US riboflavin production. The pharmaceutical grade product from chemical synthesis (Hoffmann-La Roche) sells for only marginally more ($55–60 kg^{-1}). Biologists can draw some comfort from the fact that at least the ribose is made by fermentation (by Takeda).

Riboflavin can be obtained at concentrations of 6.5 g l^{-1} in the final broth (Perlman, 1978), and some manufacturers may now achieve even more. This result has also involved a long process of strain selection and medium plus fermentation optimization. The ancestral strains produced less than 0.5 g l^{-1}. The fermentation is expensive because maximal synthesis can only be obtained in a complex broth containing glucose, corn steep liquor, meat scraps, peptone, enzymically degraded collagen and lipids plus yeast extract. It is also a lengthy (several days) batch operation, because vitamin over-production occurs after growth has ceased. Coupled with a complex and costly extraction procedure the fermentation process cannot possibly compete in the pharmaceutical grade market. Riboflavin was formerly also obtained from cells of *Cl. acetobutylicum* as a by-product of the acetone/butanol fermentation.

A similar situation would undoubtedly exist with vitamin B$_{12}$ (cyanocobalamin), but this molecule is too complex to be synthesized chemically. Instead it is produced exclusively by fermentation. Despite some improvements in yield the maximum final broth concentrations are still in the 30–80 mg l^{-1} range (Table 5.1). It is entirely intracellular, so cells are separated from the broth and the vitamin obtained by a lengthy procedure involving solvent extractions and subsequent purification (Florent & Ninet, 1979). As a result of the yields, concentration and extraction, vitamin B$_{12}$ sells for around $8 per g (the price of gold in August 1984 was $12 per g). A cheaper animal feed product is obtained by drying whole cells, as with riboflavin. World production of vitamin B$_{12}$ is static at 9–10 te per year of which 6 te is purified, the remainder being animal feed. Most manufacturers are pharmaceutical companies such as Glaxo (UK), Merck (US) and Rhone Poulenc (France).

Survey of fermentation processes

The list of examples here is not exhaustive and contains some notable omissions, but it includes most of the significant products in terms of current sales values. Traditional fermented foods and beverages are too numerous and diverse to include in this work. Some products have not been included because of relatively small volumes and lack of available information, for example dihydroxyacetone and nucleotides. Enzymes are omitted because of the difficulty of quantifying yields on the basis of activity. Some aspects of their production are described in Chapter 7.

The most important group not covered here are the antibiotics other than penicillin, notably aminoglycosides, cephalosporins and tetracyclines. They have very significant sales values of *c.* $4 billion but very few data on their production costs are available. In general it is fair to say that most lie between penicillin and vitamin B$_{12}$ in the cost spectrum, although bulk tetracycline is approximately the same price as bulk penicillin. Generally they are more expensive than penicillin to produce because of lower conversion efficiencies, broth concentrations and more expensive recovery processes. Some have sales prices as high as vitamin B_{12}, but these values are distorted by factors in pharmaceutical pricing (Chapter 2) and because many have a number of expensive chemical synthetic steps after the fermentation stage.

Bearing these omissions in mind it is worth comparing the main parameters of these products of fermentation. First in terms of total value at over $10 billion world-wide they exceed all other non-traditional biotechnology processes. In turn they are dominated by the contribution of ethanol at $4 billion and antibiotics at maybe $4–5 billion (excluding chemical steps). Of the remainder, well-established processes such as citric acid and yeast production make a large contribution; these are mature products with little growth. Some of the other products, such as lactic acid and riboflavin, have suffered a decline in the face of competition from the chemical industry. Some products are growing, notably the amino acids and xanthan gum (although monosodium glutamate is probably beginning to plateau). Spectacular growth of fermentation ethanol has occurred in recent years although it is confined to Brazil and the United States which together account for 80–90% of world production. India makes a high proportion of the remainder. Apart from ethanol and lactic acid all the products shown in Table 5.12 are obtained by aerobic processes.

With the exception of ethanol and some applications of xanthan the products are all foodstuffs (or animal feed) and drugs. This affords them some protection from direct competition with the chemical industry because of toxicological regulations and natural flavour characteristics. Mostly however it is fair to say that the fermentation routes survive because a chemical synthesis, if possible at all, is more complex and more costly. In the case of drugs and to a certain extent vitamins, demand is highly inelastic so very expensive processes can survive, particularly in view of the small quantities required.

In conclusion, fermentation can be said to occupy a niche in productive industry supplying a limited range of foodstuffs and drugs in an 'open market' economy. With government intervention with respect to ethanol and single cell protein the range has been extended. The limitations of range are in part due to the expenses of fermentation processes, most of which are evident in this survey. First they are all operated in dilute aqueous solutions. The maximum final broth concentration (Table 5.12) is 120 g l^{-1} for ethanol and this can only be achieved after lengthy fermentation times; processes are generally operated more economically at 80–100 g l^{-1} (Fig. 3.11). Dilute solutions mean large reactor volumes, hence higher capital costs, and increased

Table 5.12. *Fermentation survey*

Product	Approx. selling price mid-1984 ($ per te)	Sales vol. (te p.a.)	Current sales value ($ millions)	Final broth concentration g l⁻¹	Yield from substrate (%)	Substrate as proportion of selling price
Ethanol (95%)	500	8 000 000[a]	4000	80–120	45–48	40–60
Single cell protein	NA	NA	—	20–50	50	60
Yeast biomass	1 000	450 000	450	20	50	30[b]
Citric acid	1 600	300 000	480	80–100	70–80	12–15
Lactic acid[c]	2 200	20 000	44	90–100	90	15
Monosodium glutamate	2 500	200 000	550	100	40	15
Lysine	4 000	40 000	160	75	30–40	15
Xanthan	10 000	10 000	100	25	80	10
Penicillin[c]	30 000[d]	10 000[e]	1000	30	8	44[f]
Riboflavin[c]	50 000	NA	NA	10	10	—
Vitamin B₁₂	8 000 000	6[g]	48	0.05	1	—

[a] Figure is the sum of industrial production (1983) in Brazil (70% of total), USA (20%) and India (10%).
[b] Molasses.
[c] Fermentation products only.
[d] Bulk unsterile price.
[e] Total pharmaceutical sales.
[f] Total substrates (glucose, corn steep, phenylacetate etc.) 12% from glucose alone.
[g] Pharmaceutical sales only.
NA: not available.

Table 5.13. *Cost comparisons between high and low value fermentation products*

High volume, low value product	High value, low volume product
High capital cost	High capital cost
High raw material contribution	
High conversion efficiency	Greater spread of costs over fermentation and recovery
Low recovery costs	
May be operated with relatively unskilled labour in some cases	Skilled labour required
Low profit margins	High profit margins
Few regulatory problems	High legislation, regulation and plant hygiene constraints
Low research input	High research and development costs

recovery costs, again because of volume and the energy required to separate water and product. Capital costs are also high because of materials used in construction of plant and the engineering tolerances required for asepsis. If new plant is required at a realistic rate of return on investment it can account for between 20 and 50% of the cost of the product.

Raw materials costs predominate in the low value products, although the 60–70% range figures often quoted for ethanol or single cell protein are in many cases invalid. They arise from cost projections for plants that were never constructed. In plants that have been built substrate costs have been reduced in some way, either via by-product credits or government support policies. In mid-range value products there is a greater spread of costs over fermentation and downstream processing, but substrate costs may rise again in high value products because of the very low conversion efficiencies. It is more difficult to draw any firm conclusions on the relative contribution of downstream processing or conversion efficiency to price. In low value products a high conversion efficiency and low processing costs are essential, but the theoretical metabolic conversions are often limiting, notably in biomass or ethanol production from sugars. In mid- and high value products both costs rise, but there are great individual variations resulting in very different ratios. Xanthan, in common with other polysaccharides, can be obtained at a very high conversion efficiency from glucose. The expense comes in separation from the broth and purification. With penicillin, however, an efficient countercurrent extraction system has reduced downstream costs to at least equal or probably less than substrate costs (Table 7.9). Utility costs also rise in higher value products both for broth aeration, sterilization and product recovery. Labour costs are rarely more than 10% of total costs, except perhaps in some of the

traditional baker's yeast or old organic acid plants. Modern
fermentation plants are highly automated – they are capital but not
labour intensive. In addition they all have problems of disposal of large
volumes of high BOD wastes which have not generally been included in
these cost breakdowns.

A summary of the comparative costs of high and low priced products
is given in Table 5.13. As fermentation is an expensive technology and
these expenses are limiting the application of biotechnology to broader
areas of the economy, it is important to examine where costs might be
reduced and what technology is available or is being developed to
achieve this.

Process improvements

Fermenter productivity

The first target in reducing fermentation costs is to increase the
concentration of product or increase its rate of production. The latter
will reduce cycle time in a batch fermentation or enable a continuous
fermentation to be run at a higher dilution rate. Both aims will reduce
fermenter capacity for the same volume of product and reduce capital
costs. Higher final concentrations will also reduce downstream recovery
costs. Most organisms however suffer from product inhibition. Most of
the lower value fermentations are already operated at the maximum
concentration that can be tolerated and the strains have usually been
selected for enhanced resistance over many years. In the case of
biomass production the limiting concentration of cells is determined by
oxygen transfer. Costs of aeration become so high above 50 g l^{-1} that
reduced volume and recovery costs are offset by increased aeration
costs. Similarly in the low value products the conversion of substrate to
product is usually approaching the theoretical metabolic maximum, so
there is little scope for improvement there either. Reaction rates too are
very fast initially though they generally slow with increasing product
concentration. In some cases it is more cost effective to terminate the
reaction before the maximum concentration is realized.

Both reaction rates and product inhibition are areas in which genetic
engineering can have an impact. It would appear more likely that the
impact will be felt first on higher value products and particularly on
new products. Genetic engineering promises a more rational approach
to strain improvement which should cut development times of the old
empirical selection techniques for high yielding mutants. The organisms
of the mature, low cost processes are much less likely to be improved
for some time, or at least only marginally. This is because the effects of
product inhibition in particular are often multi-component and the

reaction rates are already very rapid. In these cases other areas of fermentation technology are probably more sensitive targets.

Continuous culture

Continuous culture permits an increase in fermenter productivity by eliminating the starting and stopping of batch cultures in which no production takes place and which may constitute a significant proportion of the cycle time. This down time is necessary to empty, perhaps clean, refill, sterilize, cool and re-inoculate the fermenter. In continuous culture the reactor can be operated at constant maximum productivity with continual addition of fresh sterilized medium. It can therefore reduce the fermenter capacity necessary to achieve the same production quotas. In addition there is the possibility of finer control and greater reproducibility of optimal conditions, and less investment in downstream processing which can be controlled by the flow from the fermenter, eliminating for example holding tanks required in a batch system. In the laboratory continuous culture has proved an ideal system with which to study and control bacterial growth. Organisms can be brought to a steady state under preselected limiting conditions, the effect of alteration of one variable at a time can be studied and mathematical models in growth or production kinetics constructed and used for the development of commercial processes. Small wonder then that continuous culture has received intense study and held much promise for the improvement of fermentation economics. At present however few commercial processes are operated continuously.

The two most common drawbacks of the technique in practice are the difficulty in maintaining asepsis in large scale operations and in obtaining strains which retain their synthetic abilities over long periods of time. Most continuous operations must be planned to run for weeks if not months without interruption to give significant saving over batch systems. Asepsis, always a major problem in large scale operation, is even more so under these conditions, particularly if complex media are used. Any contaminant which can outgrow the production strain will rapidly take over the culture. In batch systems infections acquired at later stages can often be ignored (or may not even be detected) because they cannot outgrow the production strain rapidly enough. Production strains too are frequently at a growth disadvantage *because* they are good production strains. They are channelling a high proportion of available nutrients into a product which in a competitive growth situation is a direct disadvantage. They often have other genetic lesions which impair growth in order to facilitate production of the desired compound. They are therefore often rapidly outgrown by contaminants, and unstable strains which revert to a form producing less product will rapidly outgrow the competition. This is particularly pronounced in strains carrying recombinant plasmids. Laboratory techniques used to select for plasmid-maintaining organisms such as antibiotics or heavy metal resistance are often not applicable on a commercial scale. It is also very difficult to devise screens to select for stable, non-reverting, high productivity mutants. It is an area requiring some clever tricks.

In conclusion, in general the very intensity of selective pressure which has made continuous culture such a valuable academic tool has adjudicated against it in commercial situations. There are however some processes which have used it. Single cell protein is the best example. Here strains are selected for high growth rates for biomass production; they do not suffer from reversion and are less vulnerable to competition. Where substrates such as methanol are used there are fewer problems with contamination because fewer organisms are able to utilize it as compared to a naturally more abundant carbon source such as glucose. Also because single cell protein production is such a marginal operation the capital and downstream cost savings conferred by continuous culture are often essential to the economic viability of the process. Even so the engineering, physiology and process considerations in such a plant require a high degree of sophistication and expertise.

Novo use a continuous fermentation to produce their glucose isomerase from a strain of *Bacillus coagulans* (Skøt, 1983). This fermentation is run on a complex medium, but at 50 °C which reduces contamination. The production strain is stable and can be maintained in continuous culture for up to 1000 h at high dilution rates giving a 3–4 times greater productivity compared to the batch process. Substrate and recovery costs are however lower in the batch process, and problems are encountered in continuous sterilization of the complex medium, which tends to coagulate and cause fouling. The clear advantage of continuous culture in this process is that it does not result in the sporulation, lysis and release of this intracellular enzyme into the medium which occurs in batch culture. This enzyme is sold as an immobilized preparation by cross linkage to cell contents (Chapter 7). Premature lysis and loss of enzyme into the medium is therefore a considerable disadvantage.

Continuous ethanol fermentation processes are offered by a number of manufacturers with many claimed advantages over the batch systems being used predominantly. Continuous systems allow the fermentation to be maintained at high ethanol concentrations (c. 100 g l^{-1}) because the effects of ethanol inhibition can be overcome by recycling cells, a relatively cheap operation if a flocculent yeast is used. Thus in addition to reduced fermenter volumes, savings can be made on distillation cost and contamination is reduced at the higher ethanol concentrations. The disadvantages of these systems are in their degree of sophistication which imposes extra costs but wipes out the advantages to many operators. Batch ethanol processes use very basic fermenters with very little cooling and no aeration. They are easy to build, easy to use, flexible in operation and reliable. Yeast is merely added to an air-saturated mash. The continuous systems require aeration, temperature control, cooling and sensitive feedback control because if ethanol levels exceed the maximum the yeasts are killed and the fermentation goes into decline. Also with many substrates recycle leads to an increase in salts concentration with a subsequent loss of activity and lowering of sugar to ethanol conversion efficiency. Only a few large scale continuous ethanol plants are in operation. This subject has been well covered in review articles (Coote, 1983; Guidoboni, 1984).

Continuous culture is a technique with cost saving potential, but which is limited by practical considerations and must be applied carefully to any individual process. A compromise semi-continuous system with some of the productivity advantages of true continuous culture is operated by some US ethanol producers, for example A. E. Staley. A batch run is extended over several days by withdrawing broth and adding fresh substrate. A similar extended fed-batch system is used by many antibiotic producers to obtain the maximum concentration in the fermenter broth. Here a proportion of the cells of a fully developed fermentation system is discharged and replaced by concentrated nutrient. The maximum rate of antibiotic synthesis can be increased up to twofold and the synthesis maintained for a longer period. Such fermentations are unstable and are regulated by fine computer control.

Reactor design

Most of the fermentations listed in Table 5.12 are carried out in stirred tank reactors which have a number of disadvantages. They have high energy requirements for aeration and agitation, the seals around the impeller shaft are a source of contamination, they can cause sheer damage to cells and local fluctuations in dissolved oxygen and carbon dioxide tensions. The problems of stirred tank reactors also tend to increase with size. As a consequence many alternative reactor designs have been proposed, but as yet only two are used in commercial applications.

Airlift designs achieve mixing and mass transfer by forcing air bubbles through a sparger system at the base of a tower fermenter. There is no mechanical agitation, the only power requirement coming from compressors delivering air to the spargers. The ICI 1500 m³ airlift fermenter has a compressor motor rated at 11 million watts. In general they are tall, the height giving high pressures at the base to increase oxygen solubility. As the gas rises through the tower the pressure drop causes CO_2 to come out of solution. In the pressure cycle system operated by ICI, effluent gas is separated from liquid at the top and the medium is recycled via a down-loop which also contains a heat exchanger to remove the excess heat of fermentation. Airlift fermenters have low shear, low energy requirements and a simplicity of construction. There are no aseptic seals around a rotating shaft. Mixing can be poor, though, and they are probably inadequate for the viscous broths of mycelial or polysaccharide fermentations. They have found most application in single cell protein fermentations where tight margins mean the utility savings are essential and continuous culture places a high reliance on the maintenance of asepsis.

Packed bed reactors are used with immobilized cells or enzymes. They are low shear systems with good heat transfer, but are often restricted in applications by poor gas transfer. They tend to be limited to single or two stage reactions, but offer substantial cost savings through re-use of catalyst (or cells) and low reactor costs. They are often employed in wastewater treatment. The use of these systems and their savings are illustrated in Chapter 7.

A great variety of other potential reactor designs, for example fluidized bed, hollow fibres, various types of loop or cyclones etc., have only been applied on a small scale. Their commercial performance is as yet uncertain.

Temperature

Most fermentations, indeed all those covered in this chapter, operate at temperatures between 25 and 35 °C because of the physiological limits of the organism. The use of thermophiles and operation at higher temperatures offers a number of potential advantages. Reaction rates and growth rates increase. A ten degree rise in temperature increases the rate of an enzyme catalysed reaction by a factor of approximately 1.8. Increased reaction rates result in higher volumetric productivity and lower capital costs. Cooling water requirements are lower and there is less heat loss after sterilization. In some tropical and sub-tropical locations this can be critical to the economics of a process, since there is no available water at a sufficiently low temperature for all or part of the year and refrigeration costs are probably prohibitive. At higher temperatures too broth viscosity is reduced with consequent impeller power savings. Chances of contamination are also reduced because of the lower incidence of thermophiles, and in some cases non-sterile feed streams may be used.

There are often savings in the cost of downstream operations such as drying and perhaps most significantly distillation. For example if we take a large plant producing 72000 te ethanol per year of 300 working days and with an average 100 g l^{-1} final ethanol concentration:

Ethanol production per day = 240 te
Ethanol production per hour = 10 te
Broth processed per hour = 100 te
If this is at 40 °C instead of 30 °C $T = 10$ deg
Energy saving 1 000 000 kcal h^{-1}
The energy content of steam at 20 lb/in^2 = 4.78 × 10^5 kcal/te
i.e. Energy saving is approximately 2 te steam/h.
With steam at *c.* $20/te = $40 steam/h/10 deg
                = $4 per te ethanol
(or approx. 1.2 ¢ per US gal).

Thus processing costs could be reduced by 1.2 ¢ per US gal per 10 deg rise in fermentation temperature. The plant could save $216000 p.a. per 10 deg. If thermophilic ethanol fermentations could be operated at say 70 °C instead of the present 32–35 °C savings on distillation alone could be very significant.

(A similar calculation of steam requirement per te of processed broth can be used to show the effect of final ethanol concentration in broth on steam for distillation costs.)

These cost savings however will be reduced by improvements in downstream processing, e.g. efficient distillation apparatus with heat recovery. In general, though, operation at higher temperatures can offer significant cost savings advantages, which can become critical in some cases, notably where there is lack of available cooling water.

Summary

Fermentation has evolved considerably from its origins in the traditional industries, but it is still an expensive and inefficient technology which will need much improvement if biotechnology is to produce bulk chemicals or food in competition with chemistry or agriculture. In present economic circumstances it can only compete with some foods, in specialist products and with political intervention. Some cost savings can be made with continuous culture and by operation at high temperatures. In the longer term the future must lie with genetically engineered strains, but most significantly with improved reactor designs and processes permitting continuous removal of products and higher product concentrations.

Further reading

The two volumes of *Microbial Technology* and *Fermentation Technology* edited
by H. J. Peppler & D. Perlman (Academic Press) give a comprehensive
coverage of many fermentations including some foods and drinks. The
chapter by W. H. Bartholomew & H. B. Reisman is a particularly valuable
treatment of economic aspects of fermentation.
Process Biochemistry is a good source of papers covering economic descriptions
of fermentations.
The Biochemical Engineering and Biotechnology Handbook by B. Atkinson and
F. Mavituna excels here, as elsewhere in biotechnology, with a thorough
coverage of process details.

6   Downstream processing

The importance of downstream processing

The separation and isolation of products from fermenter broths has long been recognized as a technically difficult and expensive procedure. It is often argued now that downstream processing is a critical limiting factor in the commercial development of biotechnology. It is certainly true that cost saving developments are essential if microbial processes are to compete with chemistry for the production of commodity chemicals. In some cases the same or similar techniques to those used in the chemical industry can be applied to biological systems; in other situations biology presents unique problems. As with other aspects of the fermentation industry there has been little commercial incentive for research into new processes until recently. Now a plethora of potential techniques is emerging, nearly all of which are untested on a commercial scale. This is in marked contrast to the rather limited range of operations which are actually practised.

Recovery plant usually represents a major investment, and isolation costs are a substantial fraction of the cost of the product. An attempt has been made to express downstream processing costs (both capital and operating) as an approximate proportion of the selling price of some major fermentation products (Table 6.1). These estimates contain a large element of guesswork because cost breakdowns are given in the form of total utilities and total labour etc. The proportion of each attributable to downstream operations in the absence of manufacturers' own costs can only be ascribed on the basis of complexity and contribution of other costs (for example substrate). There are also differences between different authors on the same product, for example penicillin (Cooney, 1979; de Flines, 1980; Rosen, 1983). Bearing this in mind a number of points emerge on the cost contribution of downstream processing. The absolute costs rise with product price and are in fact a major contribution to that price. They also rise as a percentage of product price, though there are exceptions such as penicillin, where the development of an efficient process has reduced the cost contribution. The penicillin price does however refer to the bulk product. Clinical grade material will have a higher percentage of purification costs. Some high value fermentation products such as

127

Table 6.1. *Downstream processing costs as approximate proportions of selling prices of fermentation products*

Product	Approx. selling price (US $/te)	Downstream processing % of selling price	Principal methods employed
Ethanol (95%)	500	15	Distillation
Single cell protein	400	20	Flocculation, centrifugation, drying
Yeast biomass	1 000	20	Centrifugation, drying
Citric acid	1 600	30–40	Calcium precipitation, acidification, crystallization, drying
Monosodium glutamate	2 500	30–40	Evaporation, acidification, filtration, crystallization, decolorization
Xanthan	10 000	50	Precipitation with 2-propanol, heat treatment, drying solvent recovery by distillation
Penicillin G[a]	30 000	20–30	Countercurrent extraction, crystallization (K^+ salt), solvent washing, drying
Enzymes		60–70[b]	Extracellular – remove solids, concentrate, precipitate, dry (if solid formulation)

[a] Bulk crude Penicillin G.
[b] Approximate general estimate quoted by Rosen (1983).

riboflavin cannot compete with chemically synthesized products, in part because of the high cost of purification. At the cheaper end of the range cost constraints are such that downstream operations must be simple and cheap or the product again cannot compete in the market. Thus ethanol recovery involves just one operation – distillation – although biomass is usually collected to be sold as a by-product and spent broths must be disposed of. Biomass recovery is confined to de-watering and drying operations.

Problems of scale

The most significant characteristic of fermentation products is that they are obtained in their final form in relatively dilute aqueous solutions (Table 5.13) containing a complex mixture of components, some often

Table 6.2. *De-watering costs for SCP yeasts*

Initial solids (%)	kg water to remove	kg water per kg solids	Direct drying cost (¢ at 2¢ per kg water)	Total costs ($ per te)	Centrifuge costs ($ per te solids to 30% solids)
1	99	99	198	1980	99
2	98	49	98	980	66
5	95	19	38	380	22
30	70	2.3	4.6	46	—

Drying costs calculated on 1 te steam (at $20 te^{-1}) to remove 1 te water; centrifuge costs on electricity at 5¢ per kWh.
Adapted from Labuza (1975).

very similar in physical and chemical properties. Chemical processes are frequently multi-stage in which intermediates are isolated at each stage and final concentrations are almost invariably higher. Broths also present handling problems owing to their physical nature. They are viscous, highly non-Newtonian slurries containing compressible gelatinous solids with surface layers of polysaccharide (Atkinson & Mavituna, 1983). Many fermentation products are relatively fragile, i.e. they cannot be subjected to extremes of temperature, pH or many solvents. A consequence of low broth concentrations is that initial handling stages are very expensive because of the volumes involved. For example a broth processing cost of 0.5¢ per litre at 25 g l^{-1} product translates into $200 per ton product, which is very significant in a low price bulk commodity. This was identified as a major problem in single cell protein fermentations where maximum biomass concentrations are in the 20–50 g l^{-1} range. A single centrifugal separation stage can contribute as much as $80 per ton for bacteria (Labuza, 1975). Broths are also susceptible to contamination. The maintenance of asepsis in most processing operations is virtually impossible or prohibitively expensive. Many products may also be unstable to post-fermentation changes in pH or oxygen tension. Broths must therefore be processed rapidly and products concentrated at an early stage to overcome volume costs.

De-watering and drying operations play a dominant role in downstream processing of fermenter broths. The dependence of the cost of drying upon the product concentration is clearly shown in Table 6.2. This emphasizes not only the need to operate at maximum concentration, but also the need to substitute direct drying with cheaper methods of de-watering wherever possible.

Quality

Few chemicals are sold at 100% purity. Purification is expensive, and above approximately 99.5% the cost tends to increase exponentially with the purity. Purity specifications are set to take account of cost and benefit and products are sold in grades according to usage. Thus taking for instance organic acids, there are industrial or technical grades, food and pharmaceutical grades. The high purity grades may sell for many times the price of the bulk crude product. For example, penicillin G is quoted in the $30 to $35 per kg bracket as a bulk grade product. The bulk pharmaceutical grade product sells for another $10 to $20 per kg, but if sold in sterile ampoules for injection the cost rises substantially. In this case the cost of the ampoule plus distribution and quality control checks may well exceed the cost of the product. Thus when considering purification costs it is essential to specify the grade, volume and packaging.

There are examples of products which may be sold in the form obtained from broths with very little purification other than concentration or drying. Many enzymes, notably the detergent proteases, are sold as concentrates after cell removal with very little if any purification. Some applications may be envisaged which could use whole broths if generated *in situ* or with very low transport costs, for example microbial surfactants or xanthan for enhanced oil recovery. Other products are sold without purification, but in a formulation to enhance their efficacy. Microbial pesticides, for example *Bacillus thuringiensis*, are sold combined with agents which enable them to adhere to the surface of a leaf or cause them to disperse when added to water. Active ingredients may constitute a minor component of a marketed product and it may be to the consumer's advantage that this is so. In these cases downstream processing often only comprises de-watering or drying before mixing in formulations.

In commercial terms it is essential that manufacturers maintain a strong technical sales division to align manufacturing specifications with customers' needs. Market shares can be upset by improved user technology, which increases the advantage to be gained from using a purer starting material, or by a competitor who offers a purer product at the same price. These considerations apply across the whole range of biotechnological products from bulk chemicals to diagnostic kits.

Manufacturers in some product areas, notably foods and pharmaceuticals, also face regulatory control on methods of operation, protection of personnel, waste disposal and product quality. These controls can restrict the available techniques for processing and restrict sharing of premises by different processes, while also imposing rigorous standards of plant hygiene.

Unit operations

The recovery of products can be highly individual, being related to specific physical and chemical properties such as volatility, solubility, stability etc., but a number of procedures or unit operations are used in a high proportion of cases. The most important are filtration, centrifugation, mechanical de-watering, drying, distillation and heat transfer. They are unitary concepts because problems arising from them can be solved using similar theories and interchangeable equipment. They also have similar cost considerations. In the same way the processing and purification of groups of related compounds, including proteins, can be treated from the standpoint of unit operations.

Heat transfer is of great importance in reducing operating costs. Apart from adequate insulation, such as lagged piping, heat exchangers are used wherever possible to recover energy from, for example, cooling broths or stills to heat incoming streams to evaporators or for sterilization. Few processes, however, operate at sufficiently high temperatures to produce recoverable heat, and thus the thermal hierarchy of process steps often found in chemical plants cannot be developed. Even so there have been great advances in the efficiency of heat exchange technology in recent years which will have considerable benefits for the operation of biotechnology plants. The interested reader is referred to more specific chemical engineering texts on this subject.

Many large plants are designed to operate on steam because with its high latent heat of evaporation it is more efficient than direct gas or electrical heating. A variety of fuels, such as coal, fuel oil, or wastes, depending on local costs, may be used to generate high pressure steam at 130–160 atmospheres. This is used to drive electricity-generating turbines and the waste steam at 5–10 atmospheres pressure used for sterilization, evaporation, distillation or general heating. Smaller plants do not generate their own electricity but still use steam for most heating functions.

Removal of solids

The first step in the isolation of a fermentation product is the separation of the broth into solid (predominantly cells) and soluble fractions. Solid separation steps are also used in later stages of some purifications for recovery of precipitates. In biomass or intracellular product isolation, solid separations are the first stage of de-watering the product. Conventional filtration of broths is difficult, rotary pre-coat filters being most widely used for mycelial fungal cultures, as for example in citric acid and penicillin fermentations. Centrifugation must be used for most bacteria and yeasts, but many bacteria are small (1μm or less) and have a density close to that of water. The largest industrial centrifuges are expensive and have a relatively small hydraulic capacity. For example, it was once estimated that a 100000 te year^{-1} bacterial SCP plant would require a harvesting section with 140 large industrial centrifuges. This would have added £40 per te to the cost of the product (Gow *et al.*, 1975).

The use of centrifugation to reduce drying costs is illustrated in Table 6.2. The centrifuge costs refer to yeasts which are between 5 and 7μm in size with a density between 1.04 and 1.09 g cm^{-3}. The centrifuge cost for yeasts is less than 10¢ per kg (or $100 per te) before drying. A product discharge of 20–30% solids can be obtained. For all reasonably large scale operations (more than 100 l fermenters) it is necessary to use solids-ejecting centrifuges to limit the labour cost. Starting at 20% solids (20 g l^{-1} dry wt cells) the de-watering cost is $66 for centrifugation to 30% solids plus $46 drying, a total of $112 per te yeast biomass as opposed to $980 if drying only is used. The differences in steam consumption of different evaporators or driers must also be kept in mind here.

Bacteria with average sizes between 1 and 2 μm and densities between 1.00 and 1.03 g cm^{-3} have centrifugation costs some four-fold higher than yeasts. As a consequence most bacterial SCP plants would not be viable unless a flocculation or flotation step for cell concentration precedes centrifugation. Only some bacterial strains are flocculent, so agents, usually polyelectrolytes, are added to broths. Each fermentation broth must be tested empirically to assess the efficacy of an individual flocculent.

Mycelial fungal cultures are the easiest and cheapest to handle. They can be collected and washed in a single operation using rotary filters.

Mechanical de-watering

Drying is an expensive process in terms of both capital investment and energy costs. Centrifugation can be regarded as a form of mechanical de-watering and in the example given in Table 6.2 can be seen to effect savings over drying in concentrating biomass to 30% solids; however it can only be used on cells or precipitates. Reverse osmosis is a technique finding increasing application in biotechnology for the initial stages of product concentration. It is becoming increasingly effective, but unfortunately is an area where excessive claims of efficacy have led to some scepticism amongst users. Many product streams can now be concentrated to 20 or even 30% solids, although 10–15% may still be more of a norm. Its efficiency depends on a number of factors causing membrane blinding and on the osmotic pressure of the solution. The energy saving of reverse osmosis is offset by high capital outlays and membrane replacement costs. It is very dependent on the useful membrane life and that in turn can only be

Table 6.3. *Economics for a plant concentrating 20 m³ h⁻¹ fermentation broth from 4.5%–11.2% recoverable dissolved solids (119 m³ h⁻¹ permeate)*

Running costs: 200 days year⁻¹, 20 h day⁻¹		(DKr)
Electricity (0.6 DKr kWh⁻¹, 4.5 kWh m⁻³ permeate)		127000
Membranes (one year lifetime, 1315 DKr m⁻²)		425000
Labour (2 h day⁻¹ at 150 DKr h⁻¹)		60000
Cleaning agents	approx.	20000
Total		632000
Total running costs m⁻³ permeate		13.4 m⁻³

Investment 3.3 million DKr for a complete, fully automatic plant
Yearly saving in running costs:
47200 m³ permeate year⁻¹ @ 25 DKr m⁻³ = 1 180000 DKr

Pay-back time $\dfrac{3\,300\,000}{1\,180\,000}$ year = 2.8 years

From: W. K. Nielsen & S. Kristensen (1983). The application of membrane filtration to the concentration of fermentation broths. *Process Biochemistry*, **18** (2), 8–12.

determined empirically with each individual application. Reverse osmosis technology has developed to a point where modular units are available with sufficiently long membrane lives to compete with other methods. Reverse osmosis was developed for the de-salination of water and most applications are still in water purification. It has replaced on energy savings ion-exchange and electrodialysis for the demineralization of drinking water and can now displace techniques such as flash distillation.

An example of claimed cost savings for reverse osmosis *versus* evaporation is shown in Table 6.3. It is a system developed by De Danske Sukkerfabriker, and is used to concentrate a fermentation broth from 4.5 to 11.2% solids. They used a new thin film composite membrane. The total running costs are put at 13.4 DKr m⁻³ *versus* running costs of an evaporator which are estimated at approximately 40 DKr m⁻³ water removed. On this basis the equipment repays its capital cost in 2.8 years when operated 20 h per day for 200 days per year. The major operating cost is membrane replacement which is approximately three times the energy cost with a one year membrane lifetime. It is obviously sensitive to this life span. Use on a broth which causes increased membrane blinding will rapidly reduce the savings. Conversely the technique becomes even more attractive if improved methods of membrane construction and cheaper membranes become available.

Another example of savings using reverse osmosis is given in Chapter 10 with reference to the concentration of whey permeate. This method of increasing whey utility by production of hydrolysed syrups is more sensitive to concentration than to enzyme hydrolysis costs. Development of alternative methods of concentration such as reverse osmosis may well tip the balance of economic viability in this case.

Ultrafiltration can be distinguished somewhat arbitrarily from reverse osmosis in that the membrane is permeable to both water and soluble low molecular weight compounds. Higher molecular weight compounds are excluded and can be concentrated. Operating pressures are lower with this technique than with reverse osmosis and preconcentration to about 40 or 50% may be achieved before drying is required. This application is dependent on the comparative economics *versus* filtration, centrifugation or possibly flocculation in the concentration of higher molecular weight species. New membranes with lower or more specific molecular weight cut-offs are constantly being introduced. Operating costs are again heavily dependent upon membrane life and can only be evaluated by specific application to a separation problem.

Membrane separation techniques offer selectivity, purity of product, ease of operation and economic advantage in many applications. Their economic value will be increased still further by developments in membranes and operation to higher concentrations.

Drying

Drying is one of the most important and expensive unit operations in biotechnology. It is mostly concerned with water removal, but not necessarily. Many recovery operations involve removal of other solvents as in xanthan and penicillin processing. Other than in exceptional cases, it is more economical to remove as much water as is possible from the solid by mechanical separation (filters, centrifuges, reverse osmosis) before a thermal drying operation.

There is a range of types of drying operations with different costs and advantages. Direct driers use the heat of a gas that contacts the solid to vaporize the liquid. Spray driers are the commonest form used in biological applications. Indirect driers separate the heat transfer medium from the product to be dried, usually by a metal wall. There is a wide variety of design of indirect driers, for example rotary, drum, vacuum, or steam tube; most use steam as a source of heat. The heat requirement of driers rises at low moisture contents because of heat losses and because moisture being evaporated comes from the interstices of the solid (Fig. 6.1). Practical problems such as coagulation, lump formation and surface coating of particles also cause costs to rise at low moisture content. As a consequence indirect driers are often used as evaporators only and the product is distributed and marketed as a syrup, say at 70 or 80% solids. Providing these syrups are microbiologically stable, the additional transport costs incurred by

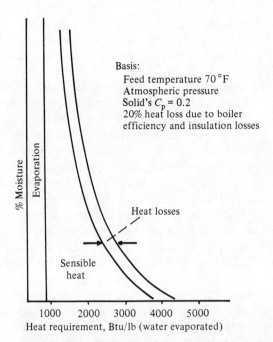

Basis:
Feed temperature 70 °F
Atmospheric pressure
Solid's $C_p = 0.2$
20% heat loss due to boiler
efficiency and insulation losses

% Moisture

Evaporation

Heat losses

Sensible
heat

1000   2000   3000   4000   5000
Heat requirement, Btu/lb (water evaporated)

Fig. 6.1. Efficiency of indirect driers. From W. L. Root (1983).
Indirect drying of solids. Excerpted by special permission from
Chemical Engineering, 2nd May, 52–6. © 1983 by McGraw-Hill,
Inc., New York, NY 10020.

the water content are less than the final drying costs. This is the case
with many sugar products such as glucose, invert and high fructose
syrups and has been proposed for whey hydrolysates. Indirect driers are
also used for solvent recovery.

Drying efficiencies are expressed in terms of water evaporated,
usually in kg steam required to evaporate 1 kg water. There is a great
range of efficiencies of different driers depending on design. Formerly
even efficient driers consumed 1.3–1.5 kg steam per kg evaporated
water, with many driers consuming over 2 kg steam. Modern two-stage
evaporators with mechanical vapour recompression use only 0.3 kg
steam per kg evaporated water and there are claims for equipment with
efficiencies as low as 0.1. There is a trade-off here of operating costs
against capital cost: all driers are expensive and the modern efficient
ones particularly so. Drying or evaporation operations can be costed
approximately by assuming the cost of steam at approximately $20 per
te and calculating the weight of water to be removed as in Table 6.2.
Here an arbitrary efficiency of 1 kg steam per kg water removed has
been assumed. In practice all driers must be tested at pilot scale on the
product removed and under the right conditions. It is still difficult to
predict the right type of drier and the effects of scale-up on factors such
as coagulation, local high temperatures and decomposition.

Spray drying is the transformation of a solution or suspension into a dry powder in a single operation. The feed liquid is atomized into a fine spray which immediately contacts a flowing stream of hot air usually at 100–200 °C. The particles are dry in less than 5 s but remain in contact with the air for 20–30 s. Spray driers are widely used in biotechnology and the dairy industry because the short contact times and relatively low temperatures are suitable for labile products. The maintenance costs are low (there are few moving parts), corrosion is low and they can be fully enclosed to minimize hazard risks. As with evaporators the performance and efficiency of spray drying plants have improved markedly in recent years. Higher inlet gas temperatures have resulted in increased efficiencies and decreased costs, but again the capital costs are relatively high. In general the efficiencies are lower than evaporators. Neglecting the inefficiency of the boiler plant and steam transfer the steam consumption of spray driers is usually in the range 1.2–1.8 kg steam per kg moisture evaporated (although many spray driers do not use steam, they are fired directly).

For many products spray drying is nevertheless expensive and may be a major cost in production. For example whey may be concentrated to 20% solids by reverse osmosis. To spray dry from this concentration entails the evaporation of 4 te water per te dry product which requires 5–7 te of steam or approximately $100–140. This is an expensive recovery of a waste product. Spray driers are often used for the final stages of drying after more efficient evaporators have been used for concentration. Thus if equipment is available many manufacturers may use a three-stage process of mechanical de-watering, evaporation, and spray drying to minimize energy consumption.

Freeze drying is the most expensive de-watering technique, but it is used for drying extremely heat sensitive materials such as blood plasma, some antibiotics, vitamins, steroids and vaccines which can stand the cost. It is also used with some foods for retention of nutritional and organoleptic properties. The wet material is frozen and controlled amounts of heat are applied under high vacuum to sublime moisture from the solid (ice) phase directly to water vapour. It is expensive because heat energy is removed by refrigeration but heat is then supplied in the drying process. To remove 1 kg water by freeze drying the following energy is used:

	kCal
Freezing product ($+20$ °C to -40 °C)	120
Sublimation	700
Freezing vapour in ice condenser down to -70 °C	730
Total (kcal per kg)	1550

This compares with approximately 600 kcal/kg in a vacuum or spray drier. Energy costs alone are three times higher than spray drying, but capital costs are even higher. It is therefore only used for specialist and expensive products, even though there have been big reductions in running and capital costs in recent years.

The drying of labile products may also be achieved at low temperatures under high vacuum using microwave driers. No detailed costs of this technique are yet available; energy consumption should be lower than freeze drying but again capital costs look to be high.

Distillation

Recovery of volatile organic compounds from fermenter broths can be achieved by distillation, which because of the opportunities for heat recovery can be made reasonably energy effective. Distillation is also used in the recovery of solvents used in processing and purification. A major factor in the expense of distillation in biotechnological processes has been historical. The fermentation producers of ethanol used the techniques of the potable industry which were concerned with flavour more than cost-effective recovery. Application of techniques developed by the petrochemical industry, such as more efficient multiple heat exchange, has resulted in substantial savings in energy from 6 kg steam l^{-1} 95% ethanol (4800 kcal l^{-1}) in a traditional process of 0.95–1.92 kg steam l^{-1} 99.5% ethanol (750–1500 kcal l^{-1}) today (Keim, 1983).

The cost of distillation is dependent on the energy source used and the location. In Brazil sugar cane bagasse is used at costs only of collecting it from nearby plantations; it is a waste with only limited alternate utility. In the United States fuel oil costs, whether used directly or for steam generation, are approximately half those in Western Europe. Similarly electricity is cheaper. A rough cost guideline can be obtained from the conversion 860 kcal = 1 kWh using the prevailing price of electricity. Generally, ethanol distillation costs, apart from the bagasse examples, lie in the range \$20–50 per te (95%) for modern stills (depending on fuel costs, boiler efficiency and so on). To obtain anhydrous ethanol demands either distillation with benzene as an entrainer or water removal by adsorbents and increases costs considerably. However 95% ethanol can be used in 10% ethanol–petroleum blends without engine modification.

As in the case of evaporators and driers the reduction in energy costs in distillation is achieved at the expense of additional capital expenditure. Within limits energy required for separation by distillation is reduced by supplying more plates in the column. Assessment of the optimum combination of capital and running costs is dependent not only on current fuel prices but on their predicted trends and pay-back or return on investment criteria which may counter the potential longer term advantage of reducing the energy used.

Distillation costs per litre of ethanol decrease with increasing broth concentration, but the product inhibition results in longer fermentation times. Again a compromise is reached at 8–9% (w/v) ethanol (Fig. 3.11). Low broth concentrations in the acetone–butanol fermentation impose high distillation costs, which similarly could be reduced by more product resistant strains. Many improvements to reduce the cost of ethanol recovery from broths have been proposed. These include fermentation at higher temperatures and reduced pressures to continuously remove ethanol from the broth and overcome product inhibition, flash distillation or higher broth concentrations achieved through continuous fermentation plus cell recycle. All incur high capital costs, all have some operating disadvantages such as recompression of CO_2 in vacuum systems and all offer only marginal savings over a modern conventional still or a cheap fuel situation. Very few have been reported in commercial operation to date.

Other separations

There are few other unit operations in biotechnology, as yet, which have widespread applications on a commercial scale to enable economic analysis or generalizations on capital or operating costs to be made over a number of cases. Most operations are specific to certain products, with specialist apparatus. Nevertheless the development of some procedures has had considerable impact on the production of certain products and could have applications elsewhere.

Solvent extraction from broths after cell removal is a very cost-effective technique used in the isolation of penicillin and some steroids. The countercurrent apparatus has high capital costs, but relatively low operating costs, most concerned with solvent recovery. The use of this technique has reduced the proportional contribution of downstream processing well below the average level for products of the same value. Some other antibiotics, notably streptomycin and cephalosporin, are absorbed directly from clarified broths onto carbon or ion exchange resins.

In theory many laboratory techniques can be scaled up. In practice some will prove unsuitable for a number of reasons, while others will become cost-effective in time with increasing use. For example chromatography is now used in a large scale process, the conversion of 42% fructose syrups to 55%. High performance liquid chromatography (HPLC) has rapidly evolved into a powerful separation technique, but the resins used are too expensive for most commercial processes because of the cost of obtaining the correct particle size. Sufficient resin for many laboratory preparative columns costs thousands of dollars, and large scale columns would be orders of magnitude higher. If techniques evolve to manufacture specific particle size resins at lower costs this technique could prove invaluable, with initial application in the separation of high value products. Affinity chromatography is similarly limited by the high cost of absorbants, and electrophoresis by problems of localized heating causing thermal destruction of products. Precipitation is widely used in bulk separations with the most commonly used reagents being (in order of increasing cost): sulphuric acid, ammonium sulphate, polyethylene glycol, ethanol and carboxymethyl cellulose. Costs are incurred in recovery of the product, for example by centrifugation, in recycling the reagent (if possible) and in the need to combat problems of high viscosity, flammability and corrosion.

Protein purification

Proteins present a particular problem in downstream recovery because they are difficult to separate from one another and are liable to degradation and denaturation. Many enzyme manufacturers have sidestepped the problem by marketing impure preparations, providing they do not contain other activities which interfere with the process or reduce yields. This can be achieved by partial purification or by isolating mutants lacking enzymes which would cause problems. Early protein purifications, for example insulin from hog pancreas, relied on precipitation techniques using reagents such as acetone or ammonium sulphate. These methods were later complemented by gel filtration, cellulose and dextran ion exchange and affinity chromatography. Combinations of such techniques are still used. They tend to be multi-step and expensive (Fish & Lilly, 1984). Solvents must be recovered or discarded, salts must be discarded, and ion exchange resins are expensive and must be regenerated. Many products, for example plasma proteins, can stand these costs, but the range of protein products now available through recombinant DNA technology provides great commercial incentives to develop new low cost techniques.

The first steps in the purification of intracellular proteins, cell disruption and the removal of cell debris, still present high cost barriers. Disruption methods used include milling, homogenizers, osmotic lysis or the use of lytic enzymes. All have drawbacks. New approaches such as the use of temperature-sensitive mutants or temperature-sensitive lytic phages have been advocated. Cross-flow filtration may prove valuable for removing cell debris and other solids, but still suffers from rapid falls in permeate flow rates at present.

Aqueous two-phase extraction systems are promising for the purification of proteins from heterogeneous mixtures. They minimize activity losses and could be suited to large scale and even continuous isolation processes. They can also be used to remove cell debris, nucleic acids, polysaccharides and coloured by-products. Phase systems can replace centrifugation, filtration and precipitation. Their economic potential in enzyme recovery has been reviewed by Kroner, Hustedt & Kula (1984), who conclude that they have low labour costs, low energy demand and low investment costs relative to precipitation, filtration and centrifugation. The drawbacks include high consumption of chemicals and disposal charges for the discarded phase, since recovery by distillation, as with organic solvents, is not possible. The chemical charges place emphasis on the use of cheap materials such as polyethylene glycol ($1000–1500 te^{-1}) and salts such as potassium and sodium phosphate, magnesium and ammonium sulphate with prices in the $400–2000 te^{-1} range. Dextran on the other hand, which has good partition coefficients when used with polyethylene glycol for protein separation, is too expensive at prices of $15000–20000 te^{-1} for crude product and up to $100000 te^{-1} for pure grades. Its use can only be contemplated for very high value products unless recovery techniques can be devised. Polyethylene glycol/salt systems for the separation of proteins from cell debris have costs of $0.2–0.6 l^{-1}, which is lower than the cost of organic solvent/water systems used in chemistry. To process 1 kg cell mass with a product yield of 85–95%, 2.5–5 l of phase system is required giving costs of $0.5–3.0 kg^{-1} cell mass to which must be added costs of waste disposal. One phase can sometimes be used two or three times, but then must be discarded. Good purifications of enzymes (up to 30-fold) have been obtained in three-step systems and up to 20-fold in two steps, but the results are very dependent on the enzyme concerned and may be as low as 1.5 to two-fold for a single step. It is nevertheless a technique which may find many applications.

Waste disposal and by-product recovery

In general, fermentation by-products are sold at very low prices or disposed of as waste. Ethanol from grain processes is an exception. Riboflavin was formerly extracted from cells as a by-product of the acetone–butanol fermentation, but there are few other examples of economic utilization of many of the potentially valuable products found in microbial cells. This is largely a consequence of their low concentrations and the high cost of the extraction procedures. They are not competitive with alternative routes. Even the biomass from many plants is incinerated or dumped, although it is a high protein material with considerable potential utility. In many cases this is because manufacturers do not wish their process strains to fall into the hands of competitors. Some biomass is purchased by animal feed manufacturers or compounders, but at very low prices. Brewer's yeast is sold (in a de-bittered form) at a premium price. Higher prices could be obtained for some biomass if human food use could be achieved, but the scale of an operation would have to be enormous to justify the cost of regulatory approval (at upwards of $20 million) and of establishing the product in the market place.

Spent broth is treated as sewage. It has a much higher biological oxygen demand than average domestic sewage. A mid-size antibiotic-producing plant has been estimated to have a daily effluent production with equivalent biological oxygen demand to a town of 30 000 inhabitants. The spent broth from a 50 000 te yr^{-1} single cell protein plant or a 150 000 te yr^{-1} ethanol plant would be equivalent to the sewage from an industrial city of 300 000 inhabitants (Atkinson & Mavituna, 1983). Many plants operate anaerobic waste treatment systems, the economics of which are discussed in Chapter 10. With new reactor designs it may be possible to operate these processes with a net benefit in the form of biogas plus fertilizer. At present they mostly represent an additional process cost. Other plants discharge untreated effluents into waterways or the sea, a practice which must ultimately be curbed. The residue from ethanol fermentations using molasses as a substrate, termed condensed molasses solubles, is sold to animal feed compounders. It can represent a credit of approximately 4¢ per US gallon of ethanol (Keim, 1983).

Carbon dioxide represents another large and for the most part unrecoverable waste from fermentation plants. Often much of the carbon in the feedstock is lost as CO_2, for example 49% in the fermentation of ethanol from carbohydrate feedstocks. Some CO_2, particularly from ethanol plants, is collected and sold, for example to soft drinks manufacturers, but supply exceeds demand. There are reports of surplus CO_2 plus low grade heat from corn wet milling plants being used for hydroponic plant culture (particularly tomatoes) in adjoining greenhouses. Corn wet milling plants in general represent outstanding examples of the value of by-product utility and are discussed in Chapter 9.

Cost factors in designing a process

It is imperative in fermentation downstream operations to handle broths rapidly and to reduce the volume of the stream handled. This is because of the lability of the product and the costs of any operations, say heating or cooling, on large volumes of broth. As a consequence de-watering procedures are important. In general mechanical

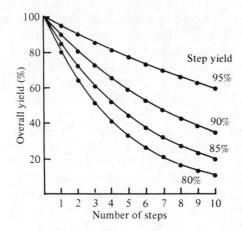

Fig. 6.2. The effects of the number of operational steps and the step yield on the overall process yield. From Fish & Lilly (1984). © 1984, Bio/Technology.

de-watering is cheaper than thermal de-watering and is used wherever possible. It may pay to use two or three different methods in sequence to reduce energy costs.

Both capital and operating costs increase with the number of steps involved in a process and where losses of yield occur at each stage the costs are amplified. The importance of the effect of step yield is shown in Fig. 6.2. There is a need either to achieve high step yields or to reduce the number of steps. This has been achieved in many established fermentations, notably antibiotics. Extracellular products require fewer steps than intracellular products and are cheaper to process.

Energy is a major component of process costs, but there is generally a trade-off with capital cost in reducing energy consumption. Later stages of purification tend to be more energy intensive than the early broth handling steps. The potential for heat recovery is generally more limited than in chemical processing. Capital costs (depreciation, return on capital and maintenance) make up a large proportion of overall costs in many operations, being 80% for rotary filtration or centrifugation and 30–55% for spray drying (Atkinson & Mavituna, 1983).

Costs increase according to the grade and purity of a product. Crude grades are often only a fraction of the cost of food or pharmaceutical grades. Many products, notably enzymes, are sold as impure products.

Plants must be designed to take account of the volumes of toxic chemicals or solvents used in a recovery process. The potential energy released by fire or explosion could be enormous. Operations cannot be transferred directly from the laboratory to manufacturing plant by simply applying scale-up factors. Pilot plant development is essential.

As elsewhere the cost of delay to a full scale plant once built but running without production is huge, not to mention factors such as loss of customers and confidence in the manufacturer.

Summary

The options that can be considered for the recovery of a product from a biotechnological process are dictated by its physical and chemical properties and its market value. A range of sophisticated processes similar to available laboratory technology can be applied in the processing of vaccines, blood plasma products, interferons, lymphokines and other high value products. Here too volume considerations and broth concentrations are not too awesome. For example even at a concentration of 0.5 g l^{-1} a recombinant DNA product might represent 500 doses per litre (at 1 mg per dose) and 10 000 litres would yield 5 million doses. Clearly fermentation volume requirements are likely to be small and high processing margins can be tolerated. At the other end of the spectrum a bulk chemical such as ethanol must sell for around $500 per ton. The raw materials may account for half this price; when capital and other factors are considered, downstream processing must be achieved for less than $100 per ton, a daunting task when broth volumes, product concentration and waste disposal are considered. Here options are very limited indeed.

The most frequently used techniques in processing biological materials are listed in Table 6.4. The list is biased towards the bulk commodity end of the spectrum, but less information is available on what is actually practised for high value products and still less on costs. There are also global considerations in this subject. Many of the techniques listed pose operational problems in developing nations. Centrifuges for example may be avoided because of the cost and difficulty of obtaining parts and a service engineer. Some energy costs, such as that of electricity, may be many times higher than in, for example, the United States.

Two specific challenges to downstream processing loom large. One is the problem of handling huge volumes of bulk products in a cost-effective way, the other is developing new technology to handle large scale separation of protein products. Neither need be restricted to a consideration of downstream operations alone. Recombinant DNA techniques make possible higher broth concentrations through gene amplification and resistance to product. They may also be used to construct strains which export formerly intracellular products, which lyse under specified conditions or do not produce proteolytic enzymes. Reduction of downstream processing costs can in fact be one of the major benefits of recombinant DNA technology. Similarly

Table 6.4. *Review of the more widely used products recovery methods in biotechnology*

Technique	Summary
Centrifugation	High capital cost and maintenance
	Fairly high energy use
Ultrafiltration/reverse osmosis	Low energy; separation on molecular weight basis but blocking problems with membranes
Evaporation/drying	High energy costs are reduced in new equipment but at the expense of capital costs
Solvent extraction	Effective, but specific applications only
Precipitation	Limited by cost of reagent or cost of recycling it
Flocculation	Often cheap and effective where applicable
Distillation	High energy use but much can be recovered as low grade heat
Ion exchange	Good separations but expensive resins which must be recycled
Crystallization	Needs fairly pure starting stream

improvements in reactor design such as fixed or fluidized beds may permit continuous operation with high concentration product streams reducing both capital expenditure and energy costs of downstream operations. Even choices of substrate and other input materials can reduce carry over of contaminants and reduce purification costs.

On the downstream hardware side new equipment permitting new processes is emerging at a rapid rate. It is impossible to say which will succeed at present, but a high degree of overall confidence can be held in some radical advances being made. Membrane technology perhaps still holds the greatest promise despite a lack of fulfilled expectations to date. Ultrafiltration has already largely replaced the precipitation techniques formerly used for protein concentration, as for example in plasma products. Cross-flow filtration may be valuable in solids and cell removal. New technology can be expected to evolve at the high priced end of the spectrum, where the products will stand the cost, and subsequently move down. Downsteam processing is presenting important business opportunities for equipment suppliers.

Further reading

Downstream process engineering and product recovery. In *Biochemical Engineering and Biotechnology Handbook*, (1983), B. Atkinson & F. Mavituna, pp. 890–931. London: Macmillan.
Fermentation and Biochemical Engineering Handbook. (1983). H. C. Vogel. Park Ridge NJ, USA: Noyes Publications.

General procedures for isolation of fermentation products. P. A. Belter, in *Microbial Technology*, 2nd edn, Vol. II (1979), ed. H. J. Peppler & D. Perlman, pp. 403–32. New York: Academic Press.

The interactions between fermentation and protein recovery. N. M. Fish & M. D. Lilly, (1984), *Biotechnology*, **2**, 623–7.

Cell collection recovery and drying for SCP manufacture. T. P. Labuza, (1975), in *Single Cell Protein*, Vol. II, ed. D. I. C. Wang & S. Tannenbaum, pp. 69–104. Cambridge Mass.: MIT Press.

Industrial Drying. (1971). A. Williams-Gardner. London: George Godwin Ltd.

7 Enzyme catalysis

Introduction and background

Enzyme catalysis can be regarded as an alternative to chemical catalysis or to fermentation. Applying strict economic principles an enzyme must be able to catalyse a given conversion more cheaply, but in practice there are areas where the three methods do not compete. Enzymes may assist detergents in cleansing formulations but they do not effectively compete with either chemical catalysis *per se* or fermentation in this application. Similarly there are many areas of chemical technology where no one can envisage enzyme applications as yet. There are however many areas where the three approaches can be regarded as competitors in principle at least, for example starch degradation, sucrose inversion, steroid conversions and vitamin synthesis.

In common with fermentation, the industrial use of enzymes has its origins in craft industries such as cheese production, textile de-sizing, tanning and juice extraction. In the absence of fundamental knowledge of the processes development has been by empiricism. The Japanese first put enzyme use on a rational footing when they discovered the active agents in the fungal breakdown of rice starch. The extraction and marketing of these enzymes as Takadiastase in the early part of this century marks the origins of the present day enzyme industry. Growth has been slow however and only in the last two decades has a sizeable growth in the industrial usage of enzymes been witnessed. Today the use of cell-free catalysis is still less than that of fermentation and much smaller than that of chemical catalysis. The gross sales of the enzyme industry, estimated at approximately $300 million world-wide in 1979/1980, have been growing at approximately 9% p.a. and were estimated at $390 million in 1983 (Hepner & Associates, 1983).

In comparison sales of chemical catalysts were $1.3 billion in the US alone in 1983 and probably in the $3–4 billion range world-wide. The principal users of these catalysts are petroleum refining, the chemical industry and automotive and industrial emissions control (pollution legislation in the US). Petroleum refining accounts for $500 million of the $1.3 billion US market, but because of the contribution of low price commodity sulphuric and hydrofluoric acid alkylation catalysts it accounts for 95% of the volume by weight (Stenson, 1983). In value

146

terms the American chemical industry uses $500 million and emissions $300 million. There is little overlap in applications between biological and chemical catalysis and direct competition between the two in these areas is unlikely. In general, competition can only be viewed in overall process terms of which catalysis is only one input. In more limited applications enzymes may replace chemical catalysis on the basis of stereo-specificity, for example in amino acid synthesis or hydroxylations of steroids. In some complex reaction sequences in the synthesis of high value compounds some steps may be performed by enzyme catalysis while others are chemical. This occurs in the synthesis of ascorbic acid, some steroids and the semi-synthetic penicillins.

Over 3000 enzymes have been characterized and approximately 200 have been sequenced, yet only 20 or so are used on any appreciable industrial scale. Most are obtained from microorganisms, with only a few coming from animals and plants. The plant enzymes, for example papain or ficin, are available only as largely unpurified powders or extracts. The availability of enzymes from animals (trypsins, rennets and lipases) depends on livestock slaughtered and in general cannot meet increasing demand. There is therefore increasing interest in cloning the genes which code for these enzymes into microorganisms. Even the number of microorganisms used for enzyme production is restricted by regulatory approval for food uses and incorporation into food processes. Hence the majority of industrial enzymes are isolated from less than 20 species of bacteria, fungi and yeasts.

Markets

Uses of enzymes in terms of volume and value are dominated by the starch, detergent and dairy industries, with smaller contributions from tanning, textiles and beverage clarification. The three principal market sectors account for at least 80% of industrial enzyme sales. A recent estimate has put starch conversion at 40% and detergents at 30% (Hepner & Associates, 1983). A similar breakdown is shown in Table 7.2, discussed below. A problem encountered in many estimates of enzyme markets is that sales of enzymes from animal sources (rennet and trypsin for example) are not included. Animal rennet sales for cheese manufacturing now total over $100 million. To a large extent though the growth of the enzyme industry has been a consequence of the growth in the starch and detergent markets, each currently running at around 5% p.a. in real volume terms. The other uses of enzymes in such applications as laboratory reagents, analysis and clinical diagnosis, while small in volume, are highly profitable and in fact increasing.

When the market for industrial enzymes is analysed for enzyme type (Table 7.1) it can be seen that hydrolytic enzymes used for the

Table 7.1. *Production of industrial enzymes*

Enzyme	Annual production expressed as tons of pure enzyme protein	Relative sales value
Bacillus protease	500	40
Amyloglucosidase	300	14
Bacillus amylase	300	12
Glucose isomerase	50	12
Microbial rennet	10	7
Fungal amylase	10	3
Pectinases	10	10
Fungal protease	10	1
Others	—	1

From: K. Aunstrup, O. Andresen, E. A. Falch & T. K. Nielsen (1979). Production of microbial enzymes. In *Microbial Technology*, 2nd edn, Vol. 1, ed. H. J. Peppler & D. Perlman, pp. 281–310. New York: Academic Press.

de-polymerization of proteins and carbohydrates predominate. The bacterial proteases, amylase and amyloglucosidase, are cheaper per unit of enzyme protein and their volumes of production are larger than the others. In recent years most widely used enzymes have shown unit cost rises well below inflation and this accounts in some measure for their increased use in the food, beverage and detergent industries. Glucose isomerase, while retailing at a higher price per unit protein, is used to achieve higher volumes of substrate conversion through immobilization on a solid support and re-use.

In the starch industry enzymes have replaced acid hydrolysis on the basis of greater efficiency, better quality products and a reduction in corrosion damage. The enzymes used are thermostable and are added to substrate mashes at high temperatures. Several enzyme processes are used: α-amylase to produce maltodextrins, β-amylase for maltose syrups, and amyloglucosidase for glucose syrups. The enzymes are generally applied in combinations starting with a liquefaction followed by saccharification to a degree depending on the desired end product. Unlike glucose isomerase the hydrolytic enzymes are still used in batch reactions because (a) they are relatively cheap, (b) high concentrations of starch create fouling problems and (c) in mainstream operations problems are caused by a build-up of intermediate hydrolysis products such as isomaltose. Immobilized amyloglucosidase is used by some manufacturers to recycle oligosaccharides in waste streams. Now also developments in mainstream operations such as co-immobilization of glucose isomerase and amyloglucosidase to lower glucose concentrations and reduce side reactions are being attempted. As a consequence of immobilization glucose isomerase costs per tonne of high fructose syrup are approximately equal to saccharification costs.

The use of enzymes in detergent formulations to remove proteinaceous dirt on soiled clothes goes back to 1913 when the German chemist Otto Rohm obtained a patent for using enzymes from animal pancreas in a pre-soaking product. This product remained on the market until after World War II, but uses of enzyme formulations in this period tended to be confined to industrial markets such as the overalls of slaughterhouse workers. The real growth of the business began in the 1960s with the isolation of the alkaline thermostable proteases from *Bacillus* species. The rapid growth which led to a 70% penetration of the detergent market in Western countries suffered a severe setback in the early 1970s because of allergies developed by production workers. This was overcome by dust reduction using pelletized formulations, and sales in Europe recovered to their previous levels fairly rapidly. In the United States the penetration of enzyme formulations is still only around 10% of the total market, but it is predicted that growth in the US will be rapid and that the European levels will be achieved within a few years. The market in Europe is thought to be mature. The present world market for detergent enzymes is put at approximately $100 million with Europe accounting for 70% of sales and the United States 15%. Growth to $200–250 million by 1990 is predicted.

Enzyme use is mature in many applications, such as the use of amylases to remove starch sizes and proteases in tanning, but use in many food processing industries is increasing. Applications include clarification of wines and fruit juices (β-glucanases and dextranases), extraction of edible oils (cellulases and pectinases), retention of colour and antioxidant properties (glucose oxidase) and others such as viscosity reduction and de-bittering. The use of β-galactosidases to hydrolyse lactose in milk products is also slowly on the increase. There are two distinct market areas here: in high price human foods such as special diets or increasing the sweetness of flavoured drinks or ice cream, and in the bulk treatment of waste whey for animal feed or fermentation applications. The bulk application is very cost-sensitive and is discussed in detail in Chapter 10.

The use of enzymes for analysis, particularly in clinical applications, has also been a strong growth area. The total market for clinical tests is in the $800 to $1000 million range world-wide, but the enzymes themselves probably contribute less than 5% of the total cost. This proportion has also been decreasing in many instances because of automation, immobilization and miniaturization. A shift from manual to automated systems can decrease the consumption of enzymes by at least 75% (Naher, 1983). The enzyme in a kit may be likened to a

Table 7.2. *Estimated size of the world-wide enzymes market ($ million)*

	Novo	Gist Brocades	Others	Total
Detergents	60	45	5	110
Starch	50	30	20	100
All others	40	15	35	90
Total	150	90	60	300

From: Grieveson, Grant and Co., analysts.

transistor in a radio or a microchip in a computer: the apparatus is built around it, with much of the success being due to cost reduction through miniaturization, which in turn leads to the essential component being only a minor cost.

Manufacturers

There has been a strong tendency to oligopoly in the bulk enzyme business. The market is dominated by two companies, Novo Industri of Denmark and Gist Brocades of Holland (Table 7.2). A more recent estimate puts the Novo share at $145 million and Gist Brocades' at $105 million in a $390 million market (Hepner & Associates, 1983). Hepner also states that seven other firms have a $70 million market, while 20 remaining firms share a $70 million market. Most notable in the second group are Chr. Hansen, based in Denmark, who have nearly 50% of the rennet market, but purchase much of their enzyme supplies from Novo.

By comparison the market in enzymes for new applications, clinical diagnosis and analysis is highly competitive; entry barriers are low and the market is characterized by a large number of small firms with high rates of innovation. It is similar to the monoclonal antibody field. Competition may also be on the increase in the bulk market. Some of the large users of enzymes such as CPC (Corn Products Co.) in the corn starch industry are reported to be entering the market as producers. The success of manufacturers will probably depend on the rate and extent to which new applications and markets for enzymes are opened up. The traditional markets are often conservative and stable.

On a national basis virtually all enzyme manufacture for free markets is carried out in Western Europe, North America and Japan, with Denmark and Holland pre-eminent. Enzyme production in the USSR and China may be significant, but the products of these countries are not available on global markets.

Other enzymes are also produced internally by users and do not figure in market terms. Most notable of these is penicillin acylase, produced internally by pharmaceutical companies. Its value in terms of volume produced or substrate converted cannot be estimated accurately.

Production and pricing of enzymes

Since microbial enzymes are relatively low volume products there has not in general been any extensive development of equipment for their production. Apparatus and techniques have been adapted from the fermentation industry and are generally stirred tank fermenters with capacities between 10 and 100 m³ operated in a batch mode. As in antibiotic production this permits flexibility where a number of different products are produced. Novo have developed a continuous process for the production of glucose isomerase, but this is a larger volume product in which this company have a strong market position.

Enzymes are generally sold in liquid formulations rather than as solids. It is more convenient for the user and cheaper as extra shipping charges are not particularly significant owing to the small volumes involved. Enzymes are rarely purified; in general extracellular enzymes are preferred and downstream processing may comprise little more than removal of whole cells by centrifugation and concentration of supernatant by reverse osmosis. Spray or vacuum drying may also occasionally be used. The result is that only a small proportion of the enzyme preparation as sold is the active enzyme. It is often only in the range of 2–5% of dry solids (Godfrey & Reichelt, 1983). From the point of view of plant operation it is preferable to add enzyme at 0.05–0.1% of volume of substrate in bulk operations. For example α-amylase is added at a rate of 1 kg per te of starch rather than, say, a syringe-full to a large reactor vessel. This reduces operator errors on large scale plant.

Enzymes may be sold on a basis of units of activity per ml or mg of preparation (as is generally the case with the detergent proteases), or by the substrate converted. For example Novo sell glucose isomerase to high fructose syrup manufacturers on the basis of tons of 42% fructose syrup it will produce during its useful lifetime. The calculation of cost per weight of enzyme protein is only of value in assessing production costs, and appears high until the utility of the enzyme is taken into consideration. For example, if an enzyme is 5% of cell protein, half of the cell weight is protein and the cell mass costs $1000 per tonne to produce, the enzyme will cost $40000 per tonne. This in fact is a cheap enzyme. The enzymes produced in large tonnage volumes cost between $10000 and $100000 per tonne, with extracellular enzymes being cheaper. The catalytic efficiency of the hydrolytic enzymes is such that amylases for example only cost between $5 and $15 per tonne of starch hydrolysed. Most commercial glucose isomerases are quoted as producing 2000 to 4000 kg fructose per kg enzyme preparation in an immobilized form. Costs are probably in the range of $10–15 per tonne of high fructose syrup produced and are certainly less than $20. Similarly the alkaline proteases represent a very small fraction of the costs of household detergents.

There is sometimes strong competition however over the price of bulk enzymes because they represent a factor, however small, over which the plant operator has control. Other costs may be fixed, for example that of substrate, by commodity price levels, futures markets or government control, and reactor costs are historical.

Diagnostic enzymes are produced in kilogram volumes at costs of up to $100000 per kg. A manufacturer of a new enzyme can afford to enter the field with only a small volume and justify a high market price by innovation. The higher overheads per volume of enzyme produced can be recovered by novelty and the fact that the whole kit price is not very sensitive to enzyme price.

Cost factors in enzyme operations

Compared to many catalysts used in the chemical industry enzymes are fragile and expensive. For example Wandrey & Flaschel (1979) draw an interesting comparison between the costs of a platinum catalyst and chymotrypsin. One mol of platinum (distributed on an aluminium oxide carrier) catalyses at 75 °C the production of 14 mol min^{-1} ethane from ethylene. One mol of chymotrypsin catalyses at 25 °C the hydrolysis of 1.335 mol of phenylalanine methyl ester min^{-1}. The respective molecular weights of platinum and chymotrypsin are 195 g mol and 24800 g mol and the prices were 13.15 DM per g for platinum at a content of 99.99% and 37.84 DM per g for chymotrypsin at 85% purity (Boehringer price list, 1976). From this they conclude that for 1 DM platinum catalyses the formation of 5.5. mmol ethane min^{-1} while chymotrypsin catalyses the formation of 1.2 nmol of phenylalanine min^{-1}. This comparison is in some respects loaded against the enzyme. Were the chymotrypsin as thermostable as many industrial enzymes it too could operate at 75 °C and catalyse 23 nmol substrate per DM. Were it an industrial protease, albeit in an impure form, it might be one to two orders of magnitude cheaper bringing the relative costs much closer. On the other hand the industrial platinum catalyst will have a longer life than the enzyme and it will be possible to recycle it. Also the enzyme will probably not operate on natural proteins at the same rate as on the selected substrate.

In some ways the value of this comparison is very limited, as the uses of the two do not overlap, but it does stress the cheapness of many chemical catalysts and the importance of enzyme price. In many bulk applications, however, improved production and user technology have brought enzyme costs down to a point where they are no longer one of the more important process costs.

It is difficult to quantify the economics of enzyme use in many applications, and to relate the efficacy of the product to a measurable parameter in an assay procedure. The purchaser is often buying a complex effect not on the basis of unit activity. For example proteases have different activities towards different substrates; one enzyme may degrade gluten much more rapidly and completely than, say, casein. Thus it is particularly difficult to quantify the effect of detergent additives. In the food industry too, complex organoleptic properties of the product confer much added value. Thus European cheesemakers are reluctant to substitute microbial rennets for the traditional product from calf stomachs because of different flavour characteristics in the product, despite the microbial enzyme being more reproducible. These properties are subjective; if microbial rennets had been developed first they may well have been adjudged superior to the calf product. In this case the selling price of the calf rennet is determined to a large extent by the availability of slaughtered animals and stomachs may sell at much higher prices than other offal as a consequence. The microbial product price has been largely determined by the calf product, being pitched to undercut it by a small amount. There has not been a great deal of pressure to undersell it by a large degree because rennet costs only represent a small factor in the overall production costs of cheese. Makers of traditional higher quality cheeses will pay the higher price for rennet which probably only represents 0.5–1.0% of the cost of manufacture. In a similar manner it is difficult to generalize on many other applications such as clarification of beverages or tanning where a number of complex factors may be considered important.

Cost-sensitivity analysis can be applied on a more rigorous basis in more defined processes, notably in the starch industry, where the cost-effectiveness of the enzyme can be measured in terms of moles substrate converted or moles product formed per unit cost. Enzyme costs *per se* are only one factor in the cost of an enzyme catalysed reaction and the enzyme costs may be itemized (Table 7.3). Some factors such as side product formation and microbial contamination may render the process unworkable if not overcome. Others may be approached by sensitivity analysis.

As an example, consider reaction temperature. This is an important parameter which has the opposing effects of increasing the catalytic activity or turnover number of the enzyme by a factor of approximately

154   *Enzyme catalysis*

Table 7.3. *Factors in the cost of enzyme reactions*

Substrate
Downstream processing (recovery)
Utilities
Plant overheads
Enzyme costs:
    Enzyme production costs (or purchase price)
    Enzyme life
    Turnover number
    Equilibrium constant for reaction
    Conversion efficiency
    Side reactions
    Temperature
    Reactor costs
    Operation concentration of substrate
    Maintenance of asepsis or low bacterial count

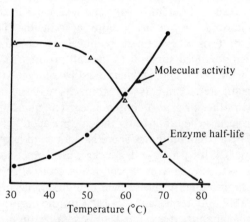

Fig. 7.1. Effect of temperature on the activity of a hypothetical
enzyme (molecular activity is number of molecules of substrate
transformed per minute per molecule of enzyme).

1.8 per 10 °C rise and reducing the useful life of the enzyme by thermal
denaturation (Fig. 7.1). In this case an optimum reaction temperature
could be selected at around 60 °C where a maximum number of moles
of substrate are converted per mole enzyme used. Again other factors
are relevant here. Should the reaction proceed more rapidly at a higher
temperature and higher enzyme cost, but lower reactor volume and
cost, because a greater throughput can be achieved with the same
equipment? The relative cost factor R is an important parameter in an
economic appraisal of an enzyme catalysed reaction. A high value of R
indicates that annual reactor cost is much more significant than the

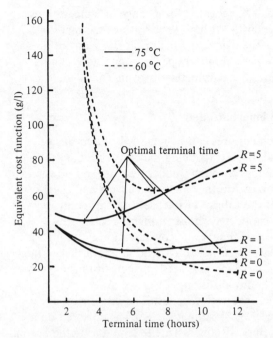

Fig. 7.2. Minimization of annual cost by slection of isothermal reaction temperature and time for glucose isomerase. From H. C. Lim & K. F. Emigholtz (1978). Optimal operations of a batch enzyme reactor: isomerization of D-glucose to D-fructose. In *Enzyme Engineering*, Vol. 3, ed. E. K. Pye and H. H. Weetall, pp. 101–14. New York: Plenum Press.

annual enzyme cost, while a value of R which approaches zero signifies that the annual enzyme cost is much more significant.

An example of selection for minimal annual costs for a batch glucose isomerase process is shown in Fig. 7.2. In this example the exact numerical value of R is not known. Costs at two reaction temperatures (60 and 75 °C) are shown for three values of R over a range of reaction times from 1–12 h. The reaction times are the times taken to reach a concentration of 42% fructose in the syrup. This analysis shows how at high R values, i.e. with relatively high reactor costs, overall costs rise steeply with time. Enzyme costs are particularly high at the lower temperature and shorter reaction times. The optimal policy in this case calls for the higher temperature and a reaction time of 4 h where enzyme and reactor costs are equivalent. A shorter reaction time is optimal at the higher temperature where reactor costs exceed enzyme costs by a factor of 5. There are distinct optimal reaction times corresponding to different R values. When the reactor cost is ignored completely ($R = 0$) time becomes irrelevant after 6–8 h.

This type of analysis can be repeated for a range of temperatures, reactor costs and enzyme costs. At high temperatures thermal denaturation will become the overriding factor, at low temperatures reactor costs will predominate. Similarly process conditions can be optimized with respect to other parameters listed in Table 7.3.

Enzyme and cell immobilization

Continuous and fed batch cultures can be regarded as a means of extending the operational life and reducing the costs of biological catalysts. Capital cost savings through smaller reactors and downstream operations are also achieved. Another method, enzyme and cell immobilization, has been the subject of intense study in recent years. A huge literature exists on support systems, coupling mechanisms, stability and reaction kinetics. Economic considerations, although of overriding significance, have received less attention. Nevertheless there is a sizeable volume of data which permits some conclusions as to the economic advantages of immobilization.

The concept is not new; invertase was adsorbed into charcoal by Michaelis & Ehrenreich (1908). A similar system was operated on a bulk scale by the British Sugar Refiners Tate & Lyle during the Second World War. They adsorbed plasmolysed yeasts containing invertase on bone char columns for the production of invert syrups from sucrose because the sulphuric acid normally used was in short supply under wartime conditions. Six large columns (6 m high and 3 m diameter) each containing 40 tons of bone char were used to decolorize and invert 25000 l of syrup per day. Bone char was chosen as a support because it was already used as a decolorizing agent in the conventional process and could be recycled by established procedures. The system was not used once the supply of acid was restored, not on cost considerations, but on quality control. The product, a high quality market leader, suffered from poorer taste qualities, too much batch-to-batch variation and microbial contamination when produced by the enzyme process. Today nearly all sucrose inversion is still carried out by acid hydrolysis.

There was a gap of over 20 years before further commercial processes utilized immobilized biological catalysts (either whole cells or cell-free enzymes). In the late 1960s and early 1970s several Japanese companies did development work on immobilized processes for the production of amino acids. Some became operational and can be used as a starting point to examine the economic advantages of enzyme and cell immobilization.

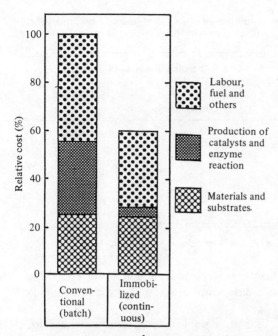

Fig. 7.3. Comparison of relative cost for industrial production of L-aspartic acid. From I. Chibata & T. Tosa (1977). Transformations of organic compounds by immobilized microbial cells. *Advances in Applied Microbiology*, **22**, 1–27.

Amino acid production

Tanabe Seiyaku Co. Ltd have been producing L-aspartic acid from fumarate and ammonia since 1973 using the enzyme aspartase contained in immobilized *Escherichia coli* cells. They claim a cost reduction of approximately 40% over the conventional batch system using a continuous immobilized process (Fig. 7.3). Materials and substrate costs remain the same, some saving (approximately 14%) is made in labour and fuel, but the majority (25%) is saved in the production of catalyst and enzyme reaction. The cells are entrapped in polyacrylamide gel and the column has a half-life of approximately 120 days at 37 °C. Unfortunately the authors do not supply additional data which would permit breakdowns in terms of fixed and variable costs, the cost of the support, concentration of streams, purification and downstream processing and so on. For example it is not possible to analyse where labour and fuel savings are made. Nevertheless it is an example of a process where catalyst costs are very significant and where they can be cut by immobilization. Since 1980 the economics of the process have been further improved by entrapping cells in

Table 7.4. *Comparative economics of conventional and immobilized cell processes for MSG ($ million)*

| | 30 million lb plants 1975 | |
	Conventional	Immobilized cells
Total fixed capital	21.0	12.9
Gross income (@ 83 c lb) (p.a.)	24.9	24.9
Production costs (p.a.)	10.5	12.5
Net income (p.a.)	14.5	12.4
Taxes (@ 48%) (p.a.)	7.0	6.0
Profit (p.a.)	7.5	6.4
Return on investment	36%	50%

From: K. Venkatasubramanian, A. Constantinides & W. R. Vieth (1978).

K-carrageenan rather than polyacrylamide, which has extended the half-life of the cells to 628 days. Tanabe are one of the three major producers of L-aspartic acid.

The second big areas of cost saving in immobilized systems is in lower reactor costs. For example potential savings in monosodium glutamate production costs have been claimed (Venkatasubramanian, Constantinides & Vieth, 1978). They are based on an economic comparison between two plants each producing 30 million lb of MSG p.a., one by batch fermentation using ten 130000 litre fermenters, the other by using cells immobilized on collagen packed in column reactors which operate continuously. Both hypothetical plants use the same carbon source, produce the same concentration of glutamic acid and have the same percentage recovery of MSG. On the basis of a three-monthly replacement of immobilized cells, overall production costs for the immobilized plant are some 20% higher because of the cost of cells and matrix, despite lower utilities, operating costs and overheads (Table 7.4). The savings in this estimate are based on replacing expensive fermenters with much cheaper column reactors, giving capital investment of $12.9 million for the immobilized plant *versus* $21.0 million for the fermentation plant. The immobilized project gave a return in investment of 50% as opposed to 36% for the conventional plant. This remains however an exercise on paper; producers of MSG still utilize batch fermentation, possibly for historical reasons, or possibly because of technical and operational problems.

Immobilized enzyme reactions can also provide significant cost reductions if the process can be carried out at high substrate and product concentrations. Savings again are made in reactor costs, which

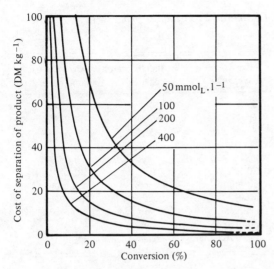

Fig. 7.4. Cost of separation per unit weight of product as a function
of conversion at different starting substrate concentrations. From
Wandrey & Flaschel (1979).

can be reduced in size for the same production levels, and in
downstream processing. In an extensive analysis of the economic
aspects of L-methionine production from *N*-acetyl-D,L-methionine by
carrier fixed acylase, Wandrey & Flaschel (1979) illustrate the
dependence of product separation costs on both conversion efficiency
and substrate concentrations (Fig. 7.4). Even at very low conversion
efficiencies of 20% the separation costs at 0.4 M substrate concentration
are lower than 100% conversion at 0.05 M, largely because less solvent
has to be evaporated. This paper contains excellent examples of the
application of cost-sensitivity analysis to an immobilized enzyme process.

On a commercial scale Tanabe Seiyaku have been using immobilized
aminoacylase to make L-valine, L-phenylalanine and L-methionine since
1969. They also have been making L-malic acid since 1980 from
fumarate using fumarase in cells of *Brevibacterium flavum* immobilized
in K-carrageenan. The continuous process produces 30 te of L-malic
acid per month using a 1000 litre column fed at a flow rate of 450 l h^{-1}
of 1 M sodium fumarate pH 7.0 (Chibata, Tosa & Takata, 1983). There
are other examples of relatively small scale operations for specialist
products where immobilized enzymes and cells have found useful
employment. There are many more where processes have been
proposed, but as yet have not been operated. In economic terms two
operations are particularly significant, the de-acylation of
benzylpenicillin to 6-aminopenicillanic acid and the isomerization of
glucose to fructose.

Table 7.5. *Comparison of batch and continuous processes for glucose isomerization*[a]

	Batch free enzyme	Batch, immob. enzyme	Continuous; immob. enzyme
Reactor volume (m³)	750	750	30
Enzyme consumption (t)	150–175	17–20	10
$MgSO_4 . 7H_2O$ consumption (t)	43	43	2.2
$CoSO_4 . 7H_2O$ consumption (t)	2.2	2.2	0
Colour formation $OD_{420\,nm}$	0.20–0.25	0.05–0.10	0–0.02
Psicose formation (%)	0.1	0.1	0.1
Product refining	Filtration, carbon treatment, cation and anion exchange	Carbon treatment, cation and anion exchange	Carbon treatment

[a] Based on monthly production of 10000 tons dry solids 42% fructose DX = 93. Activity 150 IGICU/g bulk density (IGI 350 kg/m³).

From: P. B. Poulsen & L. Zitten (1978). Novo Industri A/S studies on the isomerization of D-glucose by immobilized glucose isomerase. *Enzyme Engineering 3*, ed. E. K. Pye & H. H. Weetall, pp. 497–508. New York and London: Plenum Press.

Glucose isomerase

Production of high fructose syrups using glucose isomerase now totals nearly 4 million te p.a. with a sales value approaching $2 billion. At a rough estimate it requires 1750–2000 te of immobilized glucose isomerase. This is an example of an economic opportunity being identified and a search being made for an enzyme to exploit that opportunity, since glucose isomerases *per se* do not exist in nature (Bucke, 1983*b*). The enzymes used are xylose isomerases. They are intracellular and relatively expensive to produce, so that their utilization in processes to produce a low value product such as high fructose syrups, which must compete directly with invert and sucrose syrups, comes under severe cost constraints.

A comparison of batch and continuous systems and the immobilized and free enzymes is shown in Table 7.5. The dramatic savings in enzyme consumption and reactor volume in the continuous immobilized system are evident. As a consequence all glucose isomerase processes use continuous fixed bed column systems and are run at high substrate concentrations (*c.* 45%) so that savings in reactor costs and subsequent processing are made. The processes are also operated at high flow rates to minimize enzyme syrup contact times and reduce by-product formation. The immobilized enzyme can be used at pH 8.5 and does not require stabilization with Co^{2+} which must be removed from food grade products. There is also a lower requirement for Mg^{2+}. Colour formation is reduced in the continuous immobilized system because of the lower pH, reduced contact time and because optimum productivity can be obtained at 60–65 °C rather than 80 °C with the batch system.

The process has now acquired a high degree of sophistication and is usually computer controlled. Calculations based on mathematical models of plugged flow reactors have been made to obtain optimal enzyme performance in terms of parameters such as temperature, pH, bed height, particle size and pressure drop. As enzyme activity is lost by thermal denaturation (which is exponential with respect to time) and by poisoning by trace impurities in the substrate (which is proportional to throughput), the flow rate must be reduced progressively to compensate. Enzyme is eventually discarded when the additional reactor costs incurred by the slower flow rate exceed the cost of enzyme replacement. The developments in process technology have been reviewed comprehensively elsewhere (Antrim, Colilla & Schnyder, 1979; Hemmingsen, 1979).

The productivity of the enzyme is defined as kg fructose produced per kg enzyme during its lifetime. It is obtained by multiplying the product produced per quantity of enzyme in a given time by the life span of the enzyme. It is therefore a product of the activity and the stability of the enzyme. Manufacturers of glucose isomerase sell it to syrup manufacturers on this basis. Most first generation glucose isomerases will produce between 2000–4000 kg fructose per kg enzyme, but newer products are emerging with much higher productivities claimed by their manufacturers.

The effects of temperature on activity, stability and product formation for the Novo enzyme, currently the most widely used product, are shown in Table 7.6. This enzyme is produced by *Bacillus coagulans* in continuous culture (Chapter 5). The cells are lysed and the enzyme entrapped in beads of the lysate by cross-linking. The complex is then formed into cylinders suitable for use in column reactors with glutaraldehyde. No support or entrapping matrix is added in this case, although the major competitor's product, the Gist Brocades enzyme obtained from *Actinoplanes missouriensis*, employs gelatin cross-linked by glutaraldehyde.

It is interesting to compare the data in Table 7.6 for the immobilized enzyme with the minimization of cost by selection of reaction temperature and time for the free glucose isomerase in Fig. 7.2. In the immobilized case the importance of extended catalyst life and relatively lower reactor costs lead the manufacturers to recommend an operating temperature of 60 °C. Temperatures lower than this require larger bed volumes and result in increased viscosity of syrups and microbial contamination above acceptable levels. Above 60 °C increased enzyme

Table 7.6. *Effects of temperature on activity and product formation of immobilized glucose isomerase*

Temperature (°C)	Enzyme half-life (hours; $t_{\frac{1}{2}}$)	Design enzyme lifetime ($\times t_{\frac{1}{2}}$)	Productivity: kg dry solids (42% fructose) per kg enz. (200 IGICU^a/g)	Total enzyme bed volume (m³) for 100 tonne per day plant
65	350	2	1130	9.2
		3	1300	11.7
61	800	2	1820	12.6
		3	2100	16.1
60	1000	2	2090	13.6
		3	2430	17.4
57.5	1800	2	3100	16.2
		3	3600	20.8

^{a} IGICU = Immobilized Glucose Isomerase Column Units
= initial quantity of fructose produced by 1 g enzyme.
From: Novo Industri Data Sheets on Sweetzyme Q^R.

costs outweigh the advantages of higher throughput rates and reduced reactor costs. In the batch case the economic optimum was higher to speed throughput time in an expensive reactor.

Although the equilibrium reaction mixture contains 55% fructose the extended contact times and by-product formation make it economic to produce 42% fructose syrups, which are then enriched to 55% by chromatographic separation of some of the glucose. The 55% fructose syrups match sucrose for sweetness more exactly in carbonated acidic beverages, notably Coca-Cola and Pepsi Cola which are the major markets.

It is difficult to price conversion costs or added values in an integrated process such as corn wet milling because they are very dependent on market opportunities (Chapter 9), but in mid-1984 glucose syrup (the starting material) retailed for *c*. 13¢/lb dry solids ($286/te), 42% high fructose syrups for 20.5¢/lb dry solids ($450/te) and 55% high fructose syrups for 23.5¢/lb ($517/te). On this basis the glucose isomerization process to 42% fructose can be carried out for only $164/te dry solids (or $115/te 70% syrup).

Of this the actual conversion costs including enzyme account for less than 25% of the cost. The remainder is in pre-treatment of feedstock and product purification. The cost of enzyme alone is generally less than $15/te dry solids, but the energy costs in concentrating the final product are higher. An estimate of the major cost components in high fructose syrup production from corn, including milling and starch saccharification, is given in Table 7.7.

Table 7.7. *Major cost components*
of US high fructose syrup

Component	Production Costs, 1982 (%)
Corn	50
Energy	20
Labour	10
Chemicals	10
Enzymes	5
Miscellaneous	5
Total	100

From: C. Bucke, personal communication.

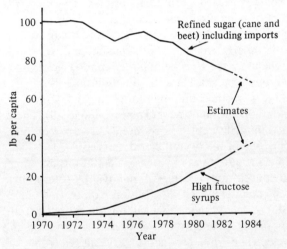

Fig. 7.5. *Per capita* sweetener consumption in the United States
1970–84. From US Department of Agriculture Statistics.

The impact of high fructose syrups on *per capita* sweetener
consumption in the United States is shown in Fig. 7.5. Since 1981 high
fructose syrup production has increased from 2.3 million te p.a. to
nearly 3.6 million te in 1984. The reduced demand is a factor in the
lowering of the world sugar price (Fig. 4.5), and it must be remembered
that the high fructose syrup process is in fact sheltered by the US
government's protection policy for indigenously produced sugar. This
in mid-1984 guaranteed a US price of \$475 te^{-1} for raw sugar (\$616 te^{-1}
for refined sugar) compared to world spot prices of \$120–140 te^{-1} for
raw sugar.

Penicillin acylase

The conversion of benzyl penicillin (penicillin G), produced by fermentation, to 6-aminopenicillanic acid is the first step in a series of reactions to make a variety of substituted penicillins which have antibiotic activity and are resistant to breakdown by β-lactamases. An enzymic process using penicillin acylase in free *E. coli* cells was originally used, but was expensive because the cells were used once and then discarded. It also suffered from carry over of material from the cells into the product stream. It was starting to be displaced by a chemical conversion, even though this was a three-stage process which had to be carried out under anhydrous conditions at a low temperature (Dunnill, 1980). Immobilization of penicillin acylase has tipped the economic balance back in favour of the enzyme route, but it is difficult to set out the economic advantages because they form part of a multi-stage synthetic route governed by internal costings. The processing lines are part of integrated production facilities which can produce other antibiotics by fermentation and often by subsequent chemical modification. Similarly production of penicillin acylase is usually performed by the user; there is not a quoted market price or market conditions determining a price as with glucose isomerase.

Bearing this in mind it is worth examining estimated capital and operating costs for 6-aminopenicillanic acid (Tables 7.8 and 7.9). These costs refer to a production facility able to produce 275 t year^{-1} of 6-APA which represents 5% of the overall world market. As described in Chapter 5 penicillin is produced in fed-batch fermentations at 25–30 g l^{-1} final broth concentration. After cell removal by rotary filters the broth is extracted by countercurrent distribution with methyl isobutyl ketone (MIBK) or butyl acetate. Penicillin is precipitated from the solvent by addition of potassium acetate, recovered by filtration and the spent solvent recycled. The crude product is recrystallized from isopropanol (IPA) which is also recycled and the product dried and packaged in bulk unsterile form or used for 6-APA manufacture. In this case costs refer to penicillin acylase immobilized on glass beads, though the matrix depends on individual manufacturers. The removal of the side chain by the enzyme liberates the free acid which is neutralized by addition of potassium hydroxide. Crude unsterile 6-APA is obtained by precipitation at pH 4.0.

The cost estimates also rely on a number of assumptions:

(1)   A 19% yield of penicillin G from fermentable sugars
(2)   Final broth concentration of 30.6 g l^{-1} from a 170 m^3 fermentation volume
(3)   A recovery of penicillin G of 85% broth total
(4)   85% conversion of penicillin G to 6-APA by a 6 h residence time in the enzyme reactors.

Table 7.8. *Estimated operating costs for 6-APA production, March 1983 Cost basis US $*

	Annual cost	Product cost (US $ kg^{-1} 6-APA)
Variable costs		
A. Raw materials		
1. Fermentation media		
Molasses	1474000	5.36
Corn steep liquor	27500	0.10
Potassium phenylacetate	937750	3.41
Potassium monohydrogen phosphate	195250	0.71
Potassium dihydrogen phosphate	233750	0.85
Other media components	9625	0.04
2. Penicillin G recovery and conversion to 6-APA		
Potassium acetate	1509750	5.49
Methyl isobutyl ketone	332750	1.21
Penicillin acylase	973500	3.54
Filter aid	324500	1.18
Other raw materials	63250	0.23
B. Utilities		
Steam	924000	3.36
Electricity	1215500	4.42
Cooling Water	101750	0.37
Total variable cost	8322875	30.27
Fixed costs		
A. Labour		
Operating labour	1727000	6.28
Maintenance labour	404250	1.47
Overheads and supervision	1067000	3.88
B. Maintenance supplies	753500	2.74
C. Taxes and insurance	503250	1.83
D. Depreciation	1677500	6.10
Total fixed cost	6132500	22.30
Total operating cost	14455375	52.57

From: F. G. Harrison & E. D. Gibson (1984). Approaches for reducing the manufacturing costs of 6-aminopenicillanic acid. *Process Biochemistry*, **19 (1)**, 33–6.

Table 7.9. *Fixed capital investment, 6-APA production facility*

Plant section	Cost (× 1000 US $)	Fraction (%)
Fermentation	7170	34.2
Clarification	1123	5.3
Extraction	895	4.3
Pen. G recovery	366	1.8
6-APA production	1045	5.0
Solvent recovery	2340	11.2
Offsites	5184	24.7
Buildings and general services	2833	13.5
Subtotal	20956	100.0
Contingency (20% of subtotal)	4191	
Total fixed capital investment	25147	

From: F. G. Harrison & E. D. Gibson (1984). Approaches for reducing the manufacturing costs of 6-aminopenicillanic acid. *Process Biochemistry*, **19** (1), 33–6.

From the capital cost schedule it can be seen that the immobilized enzyme stage is only responsible for 5% of the total compared to 34% for the fermentation facility. Equipment for broth clarification and solvent recovery are both more expensive. Similarly the cost of penicillin acylase only represents 11–12% of total variable cost or 6.7% of total operating cost (Table 7.8). Compared to glucose isomerase however the enzyme costs are still very high ($3540 per te product *versus* approximately $15). Penicillin acylases generally have an operating life of 2000–4000 h at 35–40 °C and a production of 1000–2000 kg 6-APA per kg enzyme (Godfrey, 1983). This is approximately half the productivity of glucose isomerase, but the major difference is in the production cost of the enzyme. Penicillin acylase costs $3500–$7000 kg^{-1} to produce, principally because it is produced at very low activities in cells. This emphasizes the importance of re-usability and illustrates why the batch system could not compete with a complex chemical process.

The total cost for the immobilized enzyme process has been put at 20% lower than the chemical process due to the lower capital cost, lower utility costs, a higher yield of product and re-use of the phenyl acetic acid removed in the original fermentation medium. The immobilized process is used by Beecham (UK), Bayer (US & Germany), Squibb (US) and Astra (Sweden). Different manufacturers use different supports for the enzyme. In the Astra process, for example, *E. coli* penicillin acylase is immobilized on cyanogen bromide-activated sephadex.

Other applications

The trisaccharide raffinose (galactose—glucose—fructose) is found in beet sugar liquors and inhibits sucrose crystallization. It may be hydrolysed to galactose and sucrose by α-galactosidase, but the enzyme is too expensive to be used in a single throw-away addition, as for example thermostable α-amylases are used to break down starch in cane sugar refinery streams. Immobilized enzyme column reactors are not attractive because raffinose is only found at low concentrations (3% of solids or less) in the streams, and the large volumes involved would require very large reactors. The enzyme is therefore immobilized in pellets which can be added to stirred tank reactors with sugar or molasses solutions at 30% solids in the refinery and then recovered for re-use. The enzyme is obtained from a fungal source which can be grown under conditions producing mycelial pellets and which does not need a support. It can be used in a continuous tank reactor with a residence time of 2–3 h for up to 100 days at 50 °C. Thirty kg of enzyme pellets used in this way can hydrolyse 1 ton of raffinose. In one application at the Great Western Sugar Company, Billings, Montana, USA, sucrose extraction has been increased from 87.8 to 90.7% and the capacity of the refinery has been increased by 11–12% since the introduction of the process in 1974.

Immobilized enzymes are also used widely in analytical and diagnostic kits or apparatus, often adsorbed onto glass or a membrane. Examples include the detection and estimation of glucose, urea, ethanol and acetic acid. They are frequently incorporated into an enzyme electrode, as with glucose oxidase where the H_2O_2 generated can be coupled to electron acceptors, generate a current and be converted to a digital readout corresponding to the moles of glucose present. The high cost of these enzymes makes re-use attractive. Often one immobilized preparation may serve several hundred assays.

Costs involved in enzyme immobilization

The major costs involved in the process of immobilizing an enzyme are the enzyme itself, the carrier and the immobilization procedure. An approximate estimate for fixed and variable costs involved in the immobilization of a relatively cheap enzyme on a cheap carrier (such as charcoal) in a small plant is given in Table 7.10. In this case the total production cost of £18 240 m^{-3} is comprised of 75% enzyme cost. The coupling reagents, labour, overheads and capital are relatively small, but this estimate does not take into consideration the coupling efficiency, i.e. the activity losses encountered during the procedure. This varies according to enzyme, support and procedure. For example in the Astra process for immobilizing penicillin acylase on cyanogen bromide-activated sephadex it is estimated that *c.* 50% of the

Table 7.10. *Manufacturing cost for an enzyme on a cheap carrier such as charcoal*

Item	Unit cost (£)	Cost (£)
Variable costs		
1 tonne charcoal	600/t	600
250 kg enzyme preparation	100/kg	25 000
1 250 l acetone	0.41 l^{-1}	513
34 l glutaraldehyde	1.98 l^{-1}	67
Storage drums (200 l)	13.3 each	114
		26 294

This gives 1.75 m^3 of product at a cost of approx. £15 000 m^{-3}

Fixed costs for a plant producing 50 m^3 p.a. and costing £300 000

	Cost p.a. (£)	Cost m^{-3} (£)
Depreciation of capital	30 000	600
Labour	60 000	1 200
Marketing	50 000	1 000
Services	7 000	140
Maintenance & insurance (5% of capital)	15 000	300
Total	162 000	3 240

activity is recovered on the polymer. It can be seen from this exercise that if a 50% loss of activity were obtained here the immobilized enzyme would be three times more expensive per unit activity than its free counterpart. This activity loss must be recovered in reactor costs and extended catalyst life.

The cost of the carrier can vary greatly. The example used, charcoal, is one of the cheapest. Some carriers, such as calcium alginate, widely used to trap cells in laboratory immobilization exercises, are expensive (alginate is over $3000 te^{-1}). Sephadex and derivatives are even more expensive. These supports can only be considered for high value applications, or alternatively methods must be developed for re-using the carrier. For example a manufacturer of immobilized β-galactosidases, Corning Inc., uses glass beads which are much more expensive than charcoal, but which can be re-used. Even so the Corning process is aimed at higher value milk products, as opposed to the low cost conversions aimed at upgrading whey for animal feed or fermentation applications. Some manufacturers have avoided the use of carriers altogether, as shown by the cross-linked *B. coagulans* glucose isomerase of Novo and the fungal mycelial α-galactosidase.

The contribution of carrier to enzyme immobilization costs is also dictated by the enzyme loading. High loading is necessary to reduce the carrier cost for very cheap enzymes or expensive carriers. It can also reduce contact time, reactor volume and therefore reactor costs. In turn optimum loading can only be determined with respect to other variables such as flow characteristics, column pressure drops, fouling problems and so on. Similar considerations affect the choice of carrier and are reviewed extensively elsewhere (Chibata, Tosa & Sato, 1979; Bucke, 1983*a*).

It can be appreciated that a number of important variables interact in the choice of enzyme, carrier and process which make up the costs of an immobilized enzyme system. The development of a cost-effective system can be guided by mathematical and theoretical considerations of plug flow reactors and characteristics of the materials used, but ultimately they can only be evaluated by experimental work and can only be treated in economic terms on a case by case basis. There are insufficient operating systems for more wide ranging conclusions to be made. It is important to stress however that for very cheap enzymes such as a *Bacillus* amylase the relative cost contributions of coupling and carrier are higher and the advantages of immobilization are much more marginal.

Side reactions

Although a general disadvantage of enzyme catalysis, side reactions can sometimes be amplified in immobilized reactors, where the local concentrations of reactants or products may be different from a stirred batch reactor. For example the formation of isomaltose in immobilized amyloglucosidase columns is a drawback in the use of this enzyme in mainstream starch saccharification, although it can be used to break down starch and dextrins in waste streams where the concentrations are lower.

Review of immobilization

The immobilization of enzymes and cells permits a number of cost savings and operational advantages. These vary in significance according to the process concerned, but principally involve cheaper continuous reactors and extended life of the catalyst. Many enzymes and cell systems have been shown to have greater half-lives, thermostability and other advantageous properties in their immobilized rather than free states. Immobilized systems have found three applications of major commercial significance in glucose isomerization, penicillin de-acylation and L-amino acid production, as well as a

Table 7.11. *Costs and benefits of immobilized enzymes and cells*

Advantages
Extended re-use of catalyst
Continuous process
Lower reactor costs
Enhanced stability of catalyst
Often can be used with high concentration streams – lowering recovery costs and waste disposal
Frequent enhancement of kinetic properties

Problems
Practicability in an individual application
Contribution to overall process economy
Comparatively low cost of many enzymes
New equipment required, often involving sophisticated process control
Expense of immobilization, particularly activity losses and costs of support
Side reactions
Column fouling by substrate or product streams
Microbial contamination

---

number of more minor uses in the synthesis of low tonnage speciality products or treatment of waste streams. With the notable exception of glucose isomerase their impact has not been as great or as rapid as was widely predicted a decade ago. In part, this is because of the general applications limitation of enzyme technology, but it is also because there is a trade-off between costs and benefits of immobilization. The additional costs of support and activity losses must be offset by extended use of the catalyst, lower reactor costs and other advantages (Table 7.11). Many of the variables affecting the viability of immobilized processes, such as temperature, flow rates, substrate concentration and enzyme half-life, can be analysed by cost-sensitivity techniques, but many of the important factors such as contamination and column fouling are less easy to quantify and must be tested on an empirical basis for each application.

Glucose isomerase use is as yet the only really successful large scale application of immobilized enzyme technology, but Daniels (1984) suggests that cost savings in other bulk enzyme processes such as amyloglucosidase, invertase and β-galactosidase are now possible. The significance of immobilized glucose isomerase in biotechnological terms is that it brings the technology into the range of lower cost products and lower added value processes than can be considered with other techniques such as fermentation.

Legislation

Enzymes are subject to toxicological legislation at two general levels. There is the hazard they may present to workers in the enzyme industry or to users from contact, notably allergic responses as were described with early detergent formulations, and there is also the hazard from ingestion in food and medicines. In the first case most enzymes are now marketed in solution or in pelletized form to reduce dust, and manufacturers take steps to reduce the exposure of their workers. In the second case new enzymes used in a food process must go through toxicological screening procedures in a similar way to new food additives. These procedures involve not just the enzyme, but the source organism. Thus 'natural' enzymes from the organisms already approved for food use can be cleared, but an enzyme from a non-food approved organism or a new recombinant-DNA-inserted enzyme in a food approved organism is subject to expensive and lengthy toxicological approval. This is a complex legal procedure which varies from country to country and is covered in more specialist texts (Denner, Reichelt & Farrow, 1983). The importance here is the cost. The use of several enzymes has been dropped because it was considered that the economic benefits would not repay the screening procedure. ICI in Britain have allegedly never proceeded with their genetically engineered strain of *Methlophilus methylotrophus* with *E. coli* glutamate dehydrogenase on this basis. The use of dextranases to hydrolyse contaminating dextrans in cane sugar refining is prohibited in some countries because the strains of *Penicillium* from which these enzymes are obtained can produce toxins under certain circumstances. Again the manufacturers have not considered it worthwhile to go through the procedures for the markets concerned. Genetically engineered organisms containing the calf rennet enzyme are now undergoing this procedure. The market in this case is large enough to make this worthwhile, as does the margin obtainable by using a much cheaper fermentation product process for a premium product.

It is not the intention here to pass judgements on this legislation. From the economic viewpoint it exists and must be a cost factor, often of overriding significance, to consider in developing new enzyme applications.

Summary and future prospects for enzyme catalysis

From a commercial standpoint enzymes are far from ideal. They act within very narrow temperature and pH ranges, they often have short lives and may be easily inactivated. They are fragile and expensive. The requirements for success in a commercial process are very rigorous. The

Table 7.12. *Factors affecting the economics of enzyme catalysis*

Advantages
(a) *Over chemical catalysis*
  Stereospecificity
  Mild conditions – savings in utilities

(b) *Over fermentation*
  More concentrated streams
  Cheaper recovery
  Fewer waste problems
  Cheaper reactors
  No nutritional factors
  Continuous operation easier

Problems
Thermal lability
Price
Co-factors and co-factor regeneration
Inactivity or less activity in non-aqueous solvents
Side reactions
Microbial contamination
Dilute streams relative to chemical industry

---

reaction must often be carried out at high temperatures and solids concentrations and extremes of pH to minimize contamination. The enzyme must be stable, highly active and cheap. Nevertheless enzyme catalysis does possess important advantages over the alternatives of chemical catalysis or fermentation (Table 7.12), although fermentation must be retained for complex and multi-stage processes, particularly involving co-factors.

The present bulk markets for enzymes are clearly dominated by the starch and detergent industries. The enzyme detergent market in Europe is approaching saturation at its present 70% penetration level. The United States market penetration is much lower and predicted to rise rapidly, but it too will probably plateau at a similar level. Similarly the growth of the corn starch industry is predicted to plateau over the next few years because few new large plants are being brought onstream. Glucose syrups are a mature market, and high fructose syrups will probably peak soon unless there are big developments in solid fructose as a sweetener. The potential EEC market for high fructose syrups has been blocked by legislation and it is difficult to see the need for expansion elsewhere in the world with the current oversupply of sucrose. Other older enzyme markets such as textile de-sizing and tanning are also flat. In the short term then, the enzyme industry may be predicted to expand, but in the longer term it is in need of new applications.

Several new areas and applications do hold some promise however. Genetically engineered rennets may displace the calf enzyme if they gain food approval. If they are much cheaper than the calf product the gross sales volumes may in fact fall although cheese consumption is rising by as much as 5% p.a. in some Western countries. Cheese production generates whey, one of the long term proverbial chestnuts of biotechnology research. There is still a question mark over the economics of whey utilization (discussed in Chapter 10), but in overall terms it is not particularly significant to the development of biotechnology. Enzyme routes may be used in the synthesis of Aspartame which is becoming a major new product. Many new small scale applications of enzymes in analysis and diagnostic tests are appearing, but here as in some of the bulk uses enzyme production volumes are under threat from techniques such as immobilization which increase the enzyme's life.

Even so in the longer term enzyme catalysis may be described as applications limited. This is despite the fact that hundreds of enzymes with potential applications are known. It is also in spite of the enormous amount of research which has gone into enzyme technology. It has been commented that if all the expenditure on research and development that has been done in the last 20 years were to be added up it would swamp the industry (Meers, 1983). To some extent the block may be explained by the problems listed in Table 7.12, but there are other aspects which have wider implications for biotechnology in general. Firstly, enzyme catalysis cannot be regarded in isolation, but only in the context of the entire process. For example in many cases such as starch breakdown, glucose isomerase and lactose hydrolysis, the catalytic step is not the most expensive or cost-sensitive component. Often concentration or purification costs exceed the actual catalysis costs. In other cases suitable biological raw materials, often produced by agriculture, are still more expensive than fossil fuel products. Secondly, biotechnology can make a bewildering array of new products that have no demonstrated utility. This may be a problem of marketing and gaining user acceptance. It is shared with all new technologies and often takes many years to overcome. Enzymes do not fit well into existing chemical technology, but they can do in biological systems. They will still face severe competition from chemical technology adapted to biological feedstocks, as for example is now occurring in the production of methyl glucoside from corn starch. The outcome of this competition is not clear.

In one case a clear application does exist – that of cellulose and lignin breakdown – and it is of great importance to the future development of biotechnology. The present enzyme hydrolysis is too slow and therefore reactor and other plant costs are too high for lignocellulose feedstocks to be economically viable.

Other lines of research promise to improve the commercial utility of enzymes. Genetic engineering techniques may reduce their production costs and permit expression in approved organisms. Experience gained in first generation reactors and process design will reduce costs in subsequent operations. The emerging techniques of enzyme engineering may be able to increase thermal stability or other desirable properties by inserting, deleting or changing amino acid sequences without affecting catalytic activity. Perhaps more importantly they may be able to improve catalytic activity or alter it to permit new substrates, reactions and products.

Further reading

Industrial Enzymology: The Application of Enzymes in Industry. (1983). Ed. T. Godfrey & J. Reichelt. London: Macmillan.

Microbial Enzymes and Biotechnology. (1983). Ed. W. Fogarty. London: Applied Science Publishers.

Industrial and Diagnostic Enzymes. (1983). Proceedings of a Royal Society Discussion Meeting, ed. B. S. Hartley, T. Atkinson & M. D. Lilly. The Royal Society, London.

Process development and economic aspects in enzyme engineering. Acylase L-methionine system. C. Wandrey & E. Flashel, (1979), *Adv. in Biochem. Eng.*, **12**, 147–218.

8 Energy

Introduction

Energy is one of the most important issues in economics today. In socioeconomic terms it ranks alongside population growth and food supply as a central obstacle to economic growth and social welfare. The energy problem in at least the short to medium term is an oil problem. The catalysts for the present concern and heightened awareness of energy supplies were the oil price rises of 1973–74 and 1979–80, triggered (at least) respectively by the Yom Kippur War and the Iranian Revolution. They have resulted in a ten-fold increase (four-fold in real terms) in crude oil prices in ten years and have led to projections of similar price rises interspersed with periods of adjustment (Fig. 8.1). Some authorities draw an extension of this process beyond the year 2000 with a gradual decline in oil prices as alternatives are developed. The projections are merely extensions of the events in the last decade and suffer from the limitation expressed in the rhyme: 'A trend is a trend. The problem is when will it bend? Will it go higher and higher, or just expire?' It is interesting to recall that in the early 1970s the oil price rises were not foreseen, even by experts in the oil industry or futurologists and think-tank strategists. Oil prices are now predicted to fall, but for the purpose of this chapter allowances have been made for rising prices and limited availability for at least the next two decades.

Oil has underpinned the whole world economy. The rapid growth of Western economies before 1973 was due in a large measure to cheap oil. In the US in the years 1950 to 1973 the real gross national product (i.e. the real value of goods and services) increased at a rate of 3.7% per year while energy consumption rose at 3.5% per year. During this period also the *real* price of energy declined at a rate of 1.8% per year. The impact of the oil price rises on national economies is not being overstated. If the ten years since October 1973 are compared with the ten years before it, growth rates have been cut by two-thirds while inflation and unemployment have both more than doubled. For example, while still attempting to adjust to the 1973–74 rise, the economies of OECD countries went spiralling downward during the 1979–80 rises. This second shock contributed even more to inflationary pressures, high interest rates and an increase in OECD unemployment

175

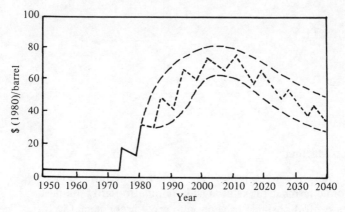

Fig. 8.1. Future oil price prediction. From A. Stratton (1981).
Decoupling the industry from oil. Reproduced with permission from
The Chemical Industry, ed. D. H. Sharpe & T. F. West, pp. 233–41.
Chichester: Ellis Horwood Ltd.

from 19 million people in 1979 to almost 32 million by the end of 1982
(Lantzke, 1982). Overall it has been estimated that the loss of income to
the OECD countries due to the second oil price rise amounted to $1000
billion in 1980 dollars. The effect has been even more devastating on
developing nations, many of whose accumulated foreign debt threatens
to break the world banking system unless they adopt policies destined
to increase unemployment and poverty. It can be argued that oil
supplies have propelled the world into an economic crisis and that
regardless of occasional upturns this impasse will continue as long as
industrial countries are susceptible to oil price rises of the magnitude
experienced in 1973–74 and 1979–80.

In the light of this situation it is hardly surprising that a profusion of
alternative energy sources to oil have been advocated. They range from
nuclear fusion and fission through alternative fossil fuels such as coal to
renewable resources such as wind, biomass and direct harnessing of
solar energy. Oil is not only important as a fuel; it is equally if not
more significant as a chemical feedstock. Biotechnological alternatives
to fossil fuels can also be used as chemical feedstocks. Before examining
potential biotechnological fuels and feedstocks it is worth briefly
outlining the nature and full extent of the oil industry. Its scale and
sophistication must be borne in mind when considering alternatives.

The oil industry

World oil consumption in 1983 was 58 million barrels per day, that is
2794 million tonnes per year. It had declined from 64 million barrels

per day in 1979, the peak year of consumption. In money terms, at approximately \$28–30 per barrel this amounts to \$1.6 billion per day or \$600 billion per year (crude prices only). To sustain a full healthy level of economic activity (if such an estimate can be made) demand would be even higher. The United States accounts for 25% of world oil consumption and Western Europe for approximately 20%. Oil tankers account for 40% of the world's shipping fleet and 50% of total tonnage. The amount of oil transported by sea rose from 250 million tonnes in 1954 to 2 billion tonnes in 1979. In the same vein, oil refineries are extremely large economic units and investment in them is huge. It has been estimated that in order to realize significant scale economies an oil refinery must be able to handle 100000 to 150000 barrels of oil per day, that is 5–7.5 million tonnes per year (Banks, 1980).

As a result the oil industry is oligopolistic. It is dominated by the seven largest companies (the 'seven sisters'). These companies are examples of vertical integration: they explore, produce, refine, transport and market oil. They in fact deal in everything from crude oil to petrochemicals. They are also extremely profitable, ranking near the top of the industrial league in terms of rate of return on equity and total capital. The twin factors of size and profitability have engendered much public hostility, but in defence of oil companies it must be said that they employ enormous technical and organizational skill and have the financial resources required to make the costly, high-risk investments on which the industry has been built (Banks, 1980).

Oil supplies are still dominated by the Organization of Petroleum Exporting Countries (OPEC), although these countries now only account for *c*. 25% of production. OPEC has been an effective cartel and price setter. It is the aim of OPEC to maximize the export price of oil and their earnings for their long term benefit. For example the aim is to develop petrochemical industries in oil-producing countries as a foundation for economic development. It makes no sense for oil-producing nations to deplete their oil reserves (capital assets) at a rapid rate, merely accumulating cash or recycling money in world capital markets for it to be eroded by inflation. In the medium term the maximum rate of price increase of oil is defined by the ability of the OECD to absorb it without economic collapse. In the long term the price will be set by the costs of alternative energy sources.

The other important factor in the pricing of oil products, particularly fuels, is taxation. This accounts for over 50% of the retail price in many countries. It is in part necessary, considering the high social cost of automobiles (roads, policing etc.), but in addition is a big net contributor to general taxation. The tax factor is of great significance and introduces an all-powerful political dimension to the development of alternatives.

Non-biotechnological alternatives to oil

In the eyes of most pundits of energy, fuels from biomass are outsiders
with long odds in the race to succeed oil. The leading contenders are
natural gas, nuclear fission, oil shale or tar sands and gasification or
liquefaction of coal.

Nuclear fission in conventional reactors requires uranium. In energy
terms uranium can only be used to generate electricity. Data from
many nations suggest that at present it can do this more economically
than coal or oil, but only if the reactor design is correct. If the more
pessimistic projections on oil come to pass, the demand for uranium for
conventional reactors could present supply problems in the early part of
the twenty-first century. Uranium mining could also be controlled by a
cartel. Its potential is fairly limited. Fast breeder reactors, which are
being developed in many countries, will extend the life of radioisotope
reserves and could provide a considerable contribution to energy supply
if developed safely. Whether the public can be convinced of such safety
in the light of one or two events at power stations and nuclear fuels
reprocessing plants is another matter.

In many countries electricity generation by coal is now cheaper than
by oil. On price per total energy content basis coal can be as little as
one-third the price of crude oil (Table 8.1), but this depends on
location. Transport costs for coal are high. Coal presents pollution
problems, principally due to sulphur content and oxides of sulphur in
exhaust gases. This is a major cause of acid rain. Coal can also provide
liquid fuels and chemical feedstocks. It was the major raw material for
the European chemical industry until the end of World War II. Coal
can be gasified, a process which is not very energy efficient, but if
carried out underground causes little pollution. It can also be used to
generate methanol and it can be liquefied to oil. This process is in
operation on a commercial scale at present in South Africa which
produces 20000 barrels (bbl) of oil from coal per day. One tonne of
coal gives 2.5 bbl oil (i.e. approx. 0.34 tonnes) in a process which is
claimed to cost $25 bbl^{-1}. On these figures it is therefore competitive
with oil, although South African coal is cheap. In the United States
pilot coal liquefaction plants operated by Gulf Oil and Exxon are
producing 30000 bbl of oil per day at a cost of approximately
$30 bbl^{-1}. A large scale programme however would require a huge
increase in coal production for a significant impact on oil consumption,
which will present infrastructure problems, such as railway capacity and
construction of new mines. The actual price of this type of oil has been
predicted to exceed $50 bbl ^{1} given problems of mining, transport,
waste disposal and pollution control.

World oil reserves are doubled if oil trapped in tar sands and shale is

Table 8.1. *Some statistics on fuels*

Volumes	1 US gallon = 3.785 litres = 0.833 Imperial gallon
	1 Imperial gallon = 4.545 litres = 1.201 US gallon
	1 barrel (bbl) = 42 US gals = 35 Imp. gals = 159 litres
	1 tonne crude oil = 7.33 bbl (average)
Conversion factors	1 kWh = 3415 Btu = 860445 g cal
	1 kCal = 3.968 Btu = 4.186 kJ
	1 tonne of oil produces about 4000 units (kWh) of electricity in a modern power station

Energy contents	Fuel	Btu lb^{-1}	kJ kg^{-1}
	Crude oil	18980	44190
	Gasoline	19850	46200
	Ethanol	12800	29800
	Methanol	9600	22300
	Coal	12500	29100
	Methane	23900	55640
	Biogas	8590	20000
	Glucose	6800	15830

World prices (mid-1984)[a]		Wt/vol basis	per MJ (1000 kJ)
	Crude oil	$30 bbl^{-1} ($220 te^{-1})	0.50¢
	Coal	$40–50 te^{-1}	0.14–0.17¢
	Methanol	c. $150 te^{-1} (US)	0.67¢
	Ethanol	$550–600 te^{-1}	1.8–2.0¢

[a] These figures are only representative. Prices vary according to grade and market.

included. The capital cost of extraction plant is very high. The investment per barrel of oil extracted per day is around $20000 which contrasts to a Middle East price of $350. The tar sands figure translates to c. $10 bbl^{-1} which, when variable costs, such as labour, transport etc. are included, adds up to a total extraction cost of c. $20 bbl^{-1}. (Saudi crude costs $0.50 bbl^{-1} and North Sea crude $12 bbl^{-1}.) This means it is still a fairly marginal operation at present crude prices, unless tax incentives are offered to the oil companies. Another price increase of the magnitude of the last two would change this situation however.

The growth of the chemical industry up to the early 1970s at a rate exceeding that of gross national products was also based on the increasing supply and decreasing real price of oil in this period. The reaction to the 1973 oil price rise was that the worst effects would be absorbed by fuels. The high added value to be derived from oil-based chemical feedstocks would buffer these products from the worst effects. Subsequent events have shed doubt on this. The dependence of the

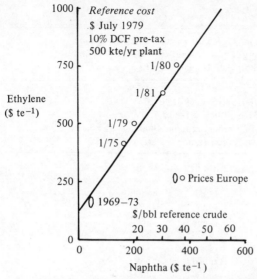

Fig. 8.2. Ethylene from naphtha. From Stratton (1981).

price of ethylene, one of the most important feedstocks, on the price of naphtha is shown in Fig. 8.2. Before 1973 the critical factor in the cost of ethylene production was the investment required in a naphtha cracker. Today the raw material is the critical cost. Also the use of coal and nuclear energy for power generation has replaced the heavy oil fractions. This has meant that the price of the lighter fractions (naphtha and gasoline) has risen faster than that of the heavier fractions such as fuel oil.

Many of the alternatives to oil are still at a serious disadvantage with present crude prices in the range $26–30 bbl^{-1}. If the price increases to the $40–60 range some will become competitive. Oil demand would then be hit by both depressed economic activity and alternative fuels. When looking at biotechnological alternatives it is reasonable to consider only those which would fall in the $30–50 or maybe $60 bbl^{-1} price range.

Biological energy resources

In the United States the contribution of biomass to energy use has been growing at a rate of 50% every five years to a point where it now contributes almost as much energy as nuclear fission or hydroelectricity (Abelson, 1983). It is way below petroleum, natural gas and coal but nonetheless above geothermal, wind and solar sources. When this biomass energy contribution is analysed however, 95% originates from combustion of wood, wood wastes and municipal plus industrial solid wastes.

The production of liquid fuels or syncrudes from wastes and biomass is possible by several methods. Simple wood distillation was used quite widely for methanol production until the 1930s. Wastes and biomass can also be converted to liquid fuels by direct hydrogenation or by first converting the feed to synthesis gas by pyrolysis, partial oxidation or steam reforming. The synthesis gas can then be converted to liquid hydrocarbons by the Fischer–Tropsch process. In this process carbon monoxide is catalytically hydrogenated to a complex mixture of aliphatic hydrocarbons similar to a high paraffin crude. These processes are outside the scope of this work.

As described in Chapter 10 low grade methane or biogas can be generated by anaerobic digestors. The gas is of very limited value as a fuel because it is only approximately 50% methane and it contains hydrogen sulphide which must be removed by scrubbers. For compression, storage and transport the carbon dioxide must also be removed. It is usually used on site for heating or steam generation or possibly to generate electricity for supply to the grid. Its potential to contribute significantly to the global energy balance is very limited in the foreseeable future in other than purely local applications.

The greatest potential for biological energy generation in the short and medium term would appear to be in liquid fuels and chemical feedstocks. Coal and uranium can already be used to generate electricity more cheaply than oil, and natural gas can be used directly for heating. Oil prices will acquire a new equilibrium as alternatives replace it, but if economic growth is to be resumed overall energy demand will rise.

Acetone and butanol

The commercial history of the acetone–butanol–ethanol fermentation goes back to the First World War when there was a high demand for acetone in the manufacture of the explosive cordite. The fermentation process was developed in Britain by Chaim Weitzmann who had previously isolated the organism *Clostridium acetobutylicum*. There was little demand for the butanol produced in this fermentation and in some cases it was stored. Many of the plants were shut down at the end of the war. In the 1920s demand grew for butanol for the synthesis of butyl acetate, used as a solvent in nitrocellulose lacquers for the automobile industry. Acetone then became more of a by-product. Fermentation plants were erected in several countries, but notably in the United States, where some 29 000 tons of butanol were being produced annually by fermentation in 1935. Corn was initially the substrate of choice, but in the 1930s molasses, a potentially cheaper source of fermentable sugars, could be utilized in a process using different *Clostridia* species.

(*a*) *From ethylene*

$$2 \; \overset{CH_2}{\underset{CH_2}{\|}} \longrightarrow 2CH_3CHO \xrightarrow{(95\% \; yield)} CH_3CHOHCH_2CHO \; (aldol)$$

(From naphtha)

$\downarrow$ Acid

$CH_3CH{=}CHCHO$ (crotonaldehyde)

$\downarrow +H_2$

$CH_3CH_2CH_2CH_2OH \xleftarrow{+H_2} CH_3CH_2CH_2CHO$ (*n*-butyraldehyde)

(*n*-Butanol)

(*b*) *From propylene* (*Oxo process*)

$$CH_2{=}CHCH_3 + 3CO + H_2O \longrightarrow CH_3CH_2CH_2CH_2OH + 2CO_2$$

Fig. 8.3. Chemical synthesis of butanol.

A chemical process for butanol synthesis from ethylene via acetaldehyde was developed in the late 1930s and a second cheaper route developed from propylene in the 1960s (Fig. 8.3). Both raw materials are obtained from naphtha. They have displaced the fermentation route to such an extent that only one commercial process has survived, that of National Chemical Products in South Africa.

Both acetone and *n*-butanol are important bulk chemical products. Butanol is used in lacquer solvents, plasticizers, hydraulic fluids, amine resins and as an intermediate in other syntheses. Acetone is used widely as a solvent and in chemical synthesis. It is therefore important to understand why the biological route has been displaced and under what circumstances it might become competitive.

As a result of the lack of operational plants it is difficult to obtain a realistic cost breakdown. An estimate is shown in Table 8.2 which assumes a yield of 40% (w/w) total solvents from fermentable sugars in molasses. In practice the operating South African plant has been reported to achieve total solvent yields of approximately 30% (Spivey, 1978). This would put molasses prices at over \$500 per te solvent, but in cane sugar growing areas the effective price of molasses can be as low as \$30 per te which would give a substrate cost of \$200 per te solvents at 30% yield.

The reasons for the poor yields are evident from the fermentation pathways (Fig. 8.4). One mole of glucose (180 g) will yield 2 moles of ethanol or 92 g. The theoretical yield is therefore 51%. In practice yields of 45–48% are generally obtained. With acetone 1 mole (58 g) is obtained per mole of glucose, a yield of 32%. A theoretical yield of 41% can be obtained with butanol. Proportions of products vary according to the organism used. For example *Cl. acetobutylicum* gives 30% acetone, 60% butanol and 10% ethanol while

Table 8.2. *Variable costs for the acetone-butanol fermentation from molasses*

The approximate requirements for 1000 kg of solvent (i.e. acetone plus butanol plus ethanol) are:

	($)
Molasses 5 te @ $80 per te	400
Water 6.5 m³ @ 2¢ per m³	0.13
Ammonium sulphate 100 kg @ $80 per te	8
Calcium carbonate 140 kg @ $50 per te	7
Steam 2.25 te at $20 per te	45
Total	460

Adapted from J. M. Paturau (1982).

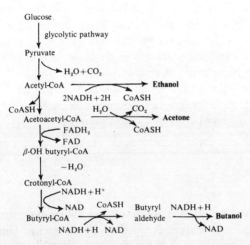

Fig. 8.4. Fermentation pathways.

Cl. saccharobutyl-acetonium liquefaciens gives 19% acetone, 78% butanol and 3% ethanol from molasses. Both organisms produce other fermentation products such as acetic and butyric acids, which explains the 30% overall yield of solvents obtained in practice.

The fermentation also suffers from low maximum concentrations in broths. *Clostridium acetobutylicum* will only tolerate 13 g l^{-1} butanol. Total solvent concentrations are usually around 20 g l^{-1}. This increases distillation costs. Operational plants have also suffered from persistent phage infections. On the plus side the spent biomass can be sold for animal feed at a premium price because it is rich in riboflavin.

Being oxidations, the chemical routes give much more favourable yields by weight – 132% butanol from ethylene and 170% from propylene – which must be borne in mind when comparing raw material costs. In practice, too, over 90% of these theoretical yields are attained. There are additional costs due to operations at high temperature and supply of hydrogen, or in the case of the propylene route, carbon monoxide, but in petrochemical complexes such materials are available at relatively low costs. Thus the petrochemical routes win out strongly on conversion efficiency from raw materials, although strictly speaking glucose cannot be considered by its weight, but by its energy value (Table 8.1). In this respect it is not so much the fermentation that is inefficient but the substrate that is expensive in terms of energy content.

A number of improvements to the fermentation process are now possible that were not available when it was last operated in many countries. As described in Chapters 5 and 6 there have been big reductions in distillation costs. The growth of the corn wet milling industry and the availability of corn starch as a product might lead to slightly lower substrate costs, but this is dubious. Genetic improvements of strains to eliminate unwanted fermentation by-products and increase thermotolerance and product tolerance could also cut costs. Perhaps the most promising area for improvement lies in the construction of strains able to ferment cheaper and more abundant substrates, notably cellulose. Many *Clostridia* are cellulose degraders and it can be predicted that the future of this fermentation must lie in the development of efficient, rapid cellulolytic and preferably thermotolerant strains.

Ethanol

Interest in the biotechnological production of fuel has been dominated by ethanol for a number of reasons. It is the cheapest product which can be produced by conventional fermentation technology because the process can be operated in simple, low cost apparatus without asepsis, and distillation is a cheap recovery process relative to others employed in the fermentation industry (Chapters 5 and 6). In addition hundreds of years of selection in the alcoholic beverage industry have produced high yielding, ethanol-tolerant, non-sporulating yeast strains.

In common with many other fermentation products interest in and development of fuel ethanol has ebbed and flowed. Internal combustion engines were first run on ethanol in Germany in 1884, but despite support from alcohol distillers and government assistance, interest had fallen by the turn of the century. After the First World War surplus alcohol was blended with petrol in the United States, but technical problems with handling were never overcome and the developments were abandoned. In Europe in the 1930s various government schemes

Table 8.3. *Chemical synthesis of ethanol*

1. Direct hydration of ethylene using phosphoric acid as a catalyst

$$CH_2{=}CH_2 + H_2O \xrightarrow{\text{H}_3\text{PO}_4} CH_3CH_2OH$$

The reaction is carried out at 68 atmospheres (1000 lb in²) and 400 °C

2. From ethylene and sulphuric acid via ethyl sulphuric acid

$$3CH_2{=}CH_2 + 2H_2SO_4 \rightarrow CH_3CH_2OSO_3H + (CH_3CH_2)_2SO_4$$
$$CH_3CH_2OSO_3H + H_2O \rightarrow CH_3CH_2OH + H_2SO_4$$

Diethyl ether is a by-product, obtained by the reaction:

$$CH_3CH_2OH + (CH_3CH_2)_2SO_4 \rightarrow (CH_3CH_2)_2O + CH_3CH_2OSO_3H$$

Table 8.4. *US alcohol production, 1977*

	Millions of gallons (anhydrous basis)		
	Beverage	Industrial	Total
Synthetic	—	244	244
Fermentation	112	84	196
Total	112	328	440

From: *Report of Alcohol Fuels Policy Review*, US Dept of Energy, Washington DC, DOE/PE-0012, June 1979, p. 111.

supported alternative fuels including methanol and ethanol, until by 1937 approximately 18% of European motor fuel was produced from wood, coal and agricultural products, even though it was grossly uncompetitive with imported petroleum (Stone, 1974; Mauldin & Phelan, 1978). In the United States various schemes were attempted, but none survived. The Second World War induced more interest, particularly in Germany, but in the era of cheap petroleum, use of fermentation ethanol virtually disappeared except in India and Brazil where it was supported by government programmes. Massive interest was of course re-awakened by the oil price rises of 1973–74.

Chemical synthesis of ethanol

Ethanol is synthesized from ethylene, in turn obtained from petroleum naphtha, by two routes (Table 8.3). The efficiency of these conversions plus low oil prices led to the displacement of fermentation ethanol from the industrial market. In the United States in the 1930s approximately 90% of industrial ethanol was supplied by fermentation, 85% using molasses as a raw material. The balance shifted dramatically in favour of the synthetic route in the 1940s and 1950s so that by 1977, 75% of

Table 8.5. *Process economics of synthetic ethanol*

	$ per US gallon	% of selling price
Variable costs		
Ethylene	1.02	55
Other materials	0.04	2
Utilities	0.27	14
Labour	0.04	2
Total variable costs	1.37	73
Fixed costs		
Plant overheads	0.03	2
Taxes and insurance	0.02	1
Depreciation	0.09	5
Total fixed costs	0.14	8
Manufacturing cost	1.51	81
Other costs		
General admin. & sales expense	0.11	6
Taxes and profit[a]		
Taxes	0.12	6.5
Net profit after tax	0.12	6.5
Assumes selling price	1.86	100

Plant capacity is 50 million gallons (150000 te) absolute ethanol per year.
Capital investment $47.2 million.
All costs refer to January 1981.
[a] Give a 25% return on investment before tax.
From: Flannery & Steinschneider (1983).

industrial alcohol was obtained from ethylene, although all beverage
alcohol is obtained by fermentation (Table 8.4). Many authorities put
the total of synthetic alcohol even higher, particularly during the 1960s.
Today there has been a dramatic change again in the United States in
favour of the fermentation product. Synthetic ethanol production was
still approximately 0.7 million te (240 million US gallons) in 1982, but
fuel ethanol production by fermentation had risen to 1.8 million te
(600 million US gallons).

The synthetic industrial product sells (absolute) for $1.7–2.0 per gallon ($0.45–0.52 per litre or $550–660 per te). It is used in the synthesis of acetaldehyde and acetic acid (43%), cosmetics and pharmaceuticals (28%), cleaning preparations and solvents (16%) and in coatings (13%). An estimate of cost breakdown is given in Table 8.5 in which ethylene costs amount to approximately $1 per gallon. One kg ethylene will yield 2.08 litres of 95% ethanol, i.e. 1.54 kg anhydrous product. (The cost dependence of ethylene on petroleum naphtha and crude oil is shown in Fig. 8.2.)

Ethanol as a chemical feedstock

Present industrial societies can be described as petroleum economies, although this is largely a post-World War II phenomenon. In 1940 95% of the 3 million tons of organic chemicals then produced came from coal and 5% from petroleum. By 1978 overall organic chemical production was increased one hundred-fold to 340 million tons, 3% of which came from coal and 97% from petroleum. Ethylene and propylene are the predominant primary building blocks for chemicals obtained from naphtha. They are made in large plants, an average modern plant producing 300000 te p.a. and big ones producing 500000 te p.a.

Ethanol can be used as a feedstock. It is possible to synthesize a wide range of compounds from ethanol by three classes of reaction: oxidation or dehydrogenation, dehydration and modification (Fig. 8.5).

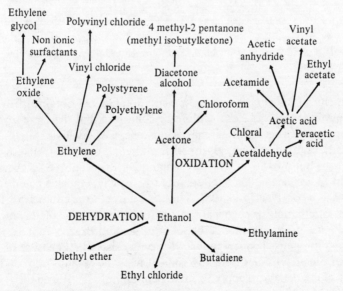

Fig. 8.5. Ethanol as a chemical feedstock.

Table 8.6. *Ethanol-based chemical industry in Brazil*

Product	Capacity (tonnes p.a.)	Uses
Ethylene	23 000	Used for low density polyethylene (closed 1969)
	10 000	Used for high density polyethylene
	4 000	Used for ethylbenzene/styrene (closed 1970)
	60 000	Ethylene dichloride (start-up 1981)
	30 000	Vinylacetate (start-up 1982)
Acetaldehyde	40 000	
	4 200	Acetic acid and solvents
	360	
	50 000	
2-Ethylhexanol	3 000	Plasticizers (currently expanding to 16 500 tonnes p.a.)
Butanol	4 800	Solvents
	150	Solvents and plasticizer (by-product of 2-ethylhexanol unit, currently expanding to 730 tonnes p.a.)
	1 530	Solvents
Butadiene	33 000	Polybutadiene (closed 1971, undergoing conversion to ethylene production)
Ethyl ether	1 400	Chemicals and pharmaceuticals
	480	Explosives
Ethylene glycol, monoethyl ether	1 300	Acetates and solvents
Diethylene glycol, monoethyl ether	1 900	Acetates and solvents
Ethyl chloride	60	Catalyst in ethylbenzene production (pilot plant)

From: M. R. Adams & G. Flynn (1982). In *Fermentation Ethanol: an Industrial Profile*, p. 10. Published by Tropical Development and Research Institute, 56/62 Gray's Inn Road, London.

A chemical industry based on fermentation ethanol obtained from renewable resources can therefore be conceived, but there are severe financial drawbacks. Brazil makes ethylene, acetaldehyde, butanol, butadiene, diethyl ether, glycol ether and ethyl chloride from ethanol at present, but mostly only in very small quantities (Table 8.6). A new plant to convert 60 000 te of ethanol to ethylene, ethylene chloride and PVC was constructed in 1981. There is a target to produce 1.5×10^9 litres of ethanol for chemical synthesis, but this is still small compared to the use of fermentation ethanol as a fuel and Brazil's rapidly expanding petrochemical industry. It only exists at all because of government subsidies. The government fixes the ethanol price at 35% of that of ethylene for supply to these plants. India developed an ethanol-based chemical industry in the 1950s, which now converts some

Table 8.7. *Possibilities of ethanol as a feedstock*

Product (plant size, t p.a.)	Production cost with ethanol		Present price with petroleum feedstock.[a] (Oil @ $28–30 bbl^{-1})
	$0.30 l^{-1}	$0.40 l^{-1}	
Acetaldehyde (45000)	0.51[b,c]	0.66	0.77
Acetic acid (60000)	0.51	0.63	0.51
Butadiene (30000)	1.14	1.45	0.77
Ethylene (60000)	0.74	0.95	0.60
2-ethylhexanol (20000)	1.35	1.66	0.84

[a] Present price refers to US bulk delivered price mid-1984 obtained from *Chemical Marketing Reporter*.
[b] All figures are in US $ per kg.
[c] Data refer to costs in cane-refining countries using molasses as a substrate. The mid-1984 ethanol price was $0.45 l^{-1}.
Adapted from J. M. Paturau (1982).

100000 tons of fermentation ethanol from molasses into PVC and polyethylene, synthetic rubber and other organic chemicals. This industry too is supported by the government to reduce petrochemical imports. The subsidy operates by giving molasses a nominally low price and blocking exports. Even so it is estimated that the PVC and polyethylene produced by this route are still more expensive than if obtained from imported petroleum.

There are two major obstacles to the development of ethanol as a feedstock; its price and the present state of technology. The direct dehydration of ethanol to ethylene using a silica–alumina catalyst at 300–360 °C requires between 1.66 and 1.81 te of ethanol to produce 1 te of ethylene. Combining this yield with the 45% fermentation yield of ethanol from sugars:

3.8 kg fermentable sugars → 1.7 kg ethanol → 1 kg ethylene

It has been estimated that ethylene will have to reach $0.80 per kg, equivalent to oil at around $40–45 per barrel, before ethylene from ethanol can become a viable proposition. Oxidation reactions are likely to achieve viability first (Table 8.7). The production costs from ethanol are optimistic and based on a location with a cheap carbohydrate feedstock. The important message is the relative favourability of the oxidation reactions.

Ethanol as a fuel

As described above, the use of ethanol as a fuel dates back to the early
days of the internal combustion engine. Today interest centres around
the use of ethanol–petroleum blends. Ethanol can be added to
petroleum at up to 20% (v/v) without modification to vehicle engines,
but anhydrous ethanol is required because water will separate out as a
layer in the base of the vehicle's petrol tank in cold weather. Modified
engines can run entirely on ethanol and in this case 95% ethanol can be
used. Although it has a substantially lower calorific value than gasoline
(Table 8.1), a greater utilization of available energy is possible from
ethanol because of higher compression ratios, better ignition and higher
burning rate. Volumetric consumption is only 15–25% higher for
straight ethanol engines on hydrated fuel (Rothman, Greenshields &
Calle, 1983) and 1–3.5% higher for 10% ethanol in petroleum blends
(average 1.8%).

Pure ethanol engines have suffered problems with cold starting, vapour
block, increased cylinder wear, engine corrosion and phase separation.
The starting problems have been combatted in Brazil by adding
gasoline to the intake air during starter operation, and the other
problems tackled by designing new engines from scratch to run on
ethanol, rather than adapting existing engines. For example the
Fiat-Uno built in Brazil to run on ethanol has a compression ratio of
10.6:1 compared with 9.2:1 in its European counterpart. The fuel lines
and carburettor have an internal coating of nickel to reduce the
corrosive effects of ethanol. These problems are not encountered in
10% ethanol–petroleum blends.

The principal marketing advantage of ethanol as a fuel additive is its
high octane number (106). It can replace tetraethyl lead as an anti-knock
agent permitting higher compression ratios without the toxicity of lead
compounds. A 10% ethanol blend raises the rating of 95 octane
petroleum to 96.7. These properties are now leading to the serious
consideration of fuel ethanol production in Western Europe as well as
in the United States, although fermentation ethanol prices are some
50% higher in Europe because of the higher prices of agricultural
commodities. Methanol can also be used as an octane boosting additive
and will be a potential rival in many countries.

Raw materials for ethanol production

It can be seen in all cost analyses that substrates are the single most
significant factor in fermentation ethanol production costs (Tables 5.5
and 5.6). There is an additional dimension to substrates when
considering the production of ethanol in the quantities required as a

Table 8.8. *Alcohol yields from various crops*

Crop	Crop yield (tonnes ha⁻¹)	Fermentable carbohydrate (% fresh wt)	Alcohol yield[a] (litres 100% ethanol per tonne)	Average yield ethanol (l ha⁻¹)
Bananas	12–50	18–20	93–104	3053
Cashew apple	—	11.6	60–68	—
Cassava	8.7	30	172–194	1592
Cocoa pulp	—	13.4	70–78	—
Maize	3.2	60	345–388	1172
Mango	—	15[b]	58–66	—
Millet	6	72	414–466	2640
Molasses (cane)	2.4–4.0	50	258–291	878
Nipa palm	—	17	88–99	—
Paddy rice	2.6	73.4–80.8	422–475	1166
Pineapples	30	14[c]	43–49	1380
Potatoes	1.6	17	98–110	166
Sorghum	1.3	68–74	391–440	540
Sugar cane	56	13–14	67–76	4032
Sweet potatoes	8.3	27[d]	154–173	1357
Sweet sorghum	25–63	13	67–76	3168
Tannia	20	17–26	98–110	2080
Taro (dasheen)	7.5–37	13–29	74–84	1738
Yam	7.5–30	15–25	86–97	1729

[a] Where a range of values for carbohydrate content is given the lesser value is used in the calculation and an 80–90% efficiency of conversion is assumed. For most cases the 80% figure can be taken as a realistic practical yield.
[b] Of mesocarp.
[c] Of edible portion which comprises 60% of whole fruit.
[d] Includes 3–6% sugar.
From: Adams & Flynn (1982).

fuel or feedstock to make any significant impact on world oil consumption: that is the type of land and the area of land required to grow crops. In Table 8.8 alcohol yields are listed from most crops considered seriously as raw materials for ethanol fermentations. In addition the alcohol yield in litres of ethanol per hectare is given. In this analysis sugar cane emerges as the clear winner, but it must be remembered that this crop requires good quality land in the tropics. Nevertheless sugar cane is the world's principal source of fermentation ethanol, usually in the form of molasses. Maize is the second most frequently used raw material, but in yields per hectare it is low in the rankings. This is because it grows in a temperate climate in countries which have large available acreages to grow it and production surplus to food and feed requirements. Its production is also highly mechanized and it can be stored. Sorghum (milo) has even poorer yields per hectare, but it will grow on marginal land. Yields are therefore only one factor in determining the suitability of a crop for ethanol production; climate, quality of land used, plus factors such as collection, seasonal availability and storage must all be considered in common with other fermentation substrates (Chapter 4).

Perhaps the most useful criterion for determining the true net cost of a raw material is the value it would have in an alternative use – its opportunity cost. For example if sawdust is considered as a substrate and it can be used in a boiler at 40% per weight of coal then the process should be charged the value of the equivalent amount of coal. This exercise becomes difficult when political barriers and food *versus* fuel uses are considered. Some nations may prefer to save fuel imports at the expense of food exports. In some cases the market values may not represent need; fuel uses may command higher prices than food. There have been unfortunate incidents of this nature already. In an effort to cut petroleum imports, Kenya started an ethanol from sugar cane programme which would have displaced food crops from high quality land and forced the nation to import food at higher prices than the petroleum it was saving. The Brazilian ethanol programme has been criticized as supplying fuel for the better off at the expense of food for the poor, but this is a more difficult claim to substantiate and is complicated by the land tenure system. This subject is beyond the scope of this work, but in general such problems are bound to arise if biotechnology is going to generate food and chemicals from agricultural products.

Cane sugar, beet sugar and molasses

Sucrose comprises up to 14% by weight of sugar cane. On the basis of typical refining yields in a country such as Brazil, and mid-1984 world market prices for raw sugar ($140 per te), the opportunity cost for sugar cane is only $7–8 per te. This cost is kept low by the world oversupply of sugar, due in part to the low rate of growth in consumption in the industrial nations, protectionist policies, dumping on world markets and the growth of high fructose syrups. In addition the fibrous residue, bagasse, left after extraction of the juice can be burned to provide heat for distillation. Brazil also controls sugar exports in order to increase supplies for fermentation.

Sugar beet, with lower yields per hectare, higher growing costs and lack of burnable residue, is not so attractive for ethanol fermentation. Substrate costs could be at least double those for cane. However there is more potential for improving the sucrose yields of the plant than there is with cane. Beet also has a harvesting season of 80 days, while cane is generally harvested for 150–200 days per year (it can only be harvested year-round in a few locations – Hawaii, Cuba and Mauritius).

The disadvantages of both sugar materials are the high cost of evaporation to concentrate the extracted juices and the rapid deterioration on storage. Plants need molasses or another substrate to run on a year-round basis.

Molasses is the most widely used substrate for ethanol production outside Brazil and the United States. Being a by-product, molasses has no production cost, the only supply factor affecting price being the transport cost from the sugar mill to the point of consumption (Chapter 4). On the demand side the only appreciable price-sensitive factor is for animal feed, but this in fact also represents the bulk of total demand (for example 83% in the US in 1978). The price fluctuations of molasses are shown in Fig. 4.6. In part because of transport costs, molasses is much cheaper in countries of origin, particularly cane growing areas, than in importing countries.

Cassava

Cassava is being developed in Brazil as a second bulk raw material for ethanol fermentation because it can grow on poorer soil than sugar cane, and cassava chips can be stored (after drying to less than 20% moisture) to use in the non-harvest season. Good conversion yields can be obtained, no acids or nutrients need be added to fermentations and deterioration of the crop is much slower than with sugar cane. There are several drawbacks, however. The lack of fibrous burnable residue means a steam requirement of 1.8 kg per kg cassava which must come from an external source. Capital costs for plant are some 20% higher than for sugar cane and a higher degree of handling of materials means higher operating costs. As a result development of cassava as a raw material for ethanol has been slower than anticipated.

Maize

The economics of maize-processing plants are discussed in detail elsewhere (Chapter 9). Only the United States uses maize as a substrate for alcohol fermentations on any scale.

Cellulose

Over the next few years fuel alcohol will undoubtedly come from cereal or sugar crops and possibly from some wastes. In view of the global abundance of cellulose and lignocellulose (Table 4.7), saccharification of these polymers must be developed for large scale fuel production. Cellulose is cheap, but it requires expensive pre-treatment, either mechanical or chemical (Chapter 10), followed by enzyme hydrolysis, which is slow and expensive in terms of enzyme and reactor costs. Also at present it is only possible to transform up to 60% of the cellulose into fermentable sugars.

Pre-treatment and hydrolysis costs of wood are similarly prohibitively expensive and the by-products, particularly hemicelluloses, are difficult to utilize. Strains of yeasts which can ferment pentoses to ethanol have been isolated, but the maximum yield is as low as 30%. Wood hydrolysis has high capital and operating costs and the hydrolysates only contain glucose at 2–4%, which means high distillation costs unless they are concentrated. Wood hydrolysis plants are only thought to operate on a commercial scale in the USSR.

The production of ethanol from both celluloses and hemicelluloses is thus technically feasible, but commercially unattractive.

Wastes

Ethanol is obtained by fermentation of whey and spent sulphite liquors from delignification processes in the production of paper pulp. Details of these wastes and other lignocellulose wastes are included in Chapter 10. Breweries often sell waste beer to local distilleries or recover the alcohol for industrial use at their own plant. It has been estimated that up to 40 million gallons of anhydrous alcohol could be produced from brewery waste streams (Lyons, 1983). A wide variety of other food wastes have been examined for ethanol production, but they all suffer from being putrescible and seasonal.

By-product credits

The sale of by-products can have a big impact on the economics of ethanol operations. Yeast biomass can be sold as animal feed, even at a premium price of up to $1000 per ton dry solids in some countries. Fusel oil, a mixture of higher alcohols collected from the rectification column, is produced at a rate of approximately 1 litre per 200 kg ethanol. It is composed mainly of amyl and isoamyl alcohols (62%), n-propanol (12%), isobutanol (15%) plus some ethanol and minor components, and is generally sold as an unrefined lacquer solvent. Its value is very small. Carbon dioxide is now used in hydroponic developments associated with big corn wet milling plants in the US. It is also sold to manufacturers of carbonated beverages, from various ethanol plants around the world, but production is far in excess of needs for this application.

In general the value of by-products with sugar cane, beet, cassava or molasses substrates is quite minor. Tops of sugar beet and cassava plants can be used as animal feed, while sugar cane bagasse has a number of applications including chipboard, paper and production of furfural, but is produced in excess by the sugar-refining industry, such that additional production is likely to be used only as a fuel. The residue after fermentation of molasses is concentrated (condensed molasses solubles) and sold as an animal feed. Its value as a by-product has been estimated at approximately 4¢ per US gal ethanol (1¢ per litre) in the United States (Keim, 1983).

Production of ethanol from corn (maize) however is completely dependent on the value of by-products. The composition of corn is approximately 72% starch, 10% protein, 4.5% oil and 13.5% cellulose, hemicellulose, ash and minor components. When corn is fermented in the whole grain or dry milling process the unfermented materials are recovered, dried and sold as distiller's dried grains and solubles (DDGS). This product is approximately 27% protein and contains reasonable levels of oil, hence it is a valuable animal feed. In the wet milling process (Chapter 9) most non-starch materials are separated from the starch before fermentation. The main products are corn oil, gluten meal (at least 60% protein) and corn gluten feed (at least 20% protein). Gluten is sold as animal feed.

The credits from the sale of by-products greatly reduce the effective raw material cost (Table 5.6). They reduce the percentage contribution of substrate to total costs of ethanol production to much lower levels than with other substrates. The values of these by-products change continually with fluctuations in the vegetable oil and animal feed markets, in particular soy bean oil and meal, their major competitors, but on average they retain a constant relationship to the price of corn. In the dry milling process, over the period 1975–81 the rate of by-product credit stayed at a constant 42% of the initial corn cost, giving an average substrate cost of $0.63 per US gallon ($0.17 per litre). Similarly the by-product credits for the wet milling process remained constant at 63% of total corn cost giving an effective raw material cost lower than that estimated in Table 5.6 of only $0.40 per US gallon ($0.11 per litre) (Keim, 1983).

The values of these credits may well be limited if volumes of ethanol production from corn keep increasing at a rapid rate. Some analysts have predicted that if US ethanol production exceeds 2 billion gallons (6 million te) *per annum* the prices of distiller's dried grains and corn gluten will decline from their present market values. Unrestricted exports of the by-product feeds would allow production to rise higher, but there are signs that the European Economic Community, one of the major markets for gluten animal feed, is considering tariff barriers. The value of by-products is so critical to the economics of the US ethanol programme that any limitation on selling will severely restrict the volume of production.

Energy balance

The return of energy obtained from 1 litre of ethanol fuel in relation to the energy expended in its production has become one of the most contentious issues in the fermentative production of fuels. It is disputed not just because of disagreement on the figures and what should be included or excluded in drawing up an energy balance, but also on its economic significance in welfare terms.

Energy balances for fermentation ethanol production can be conveniently split into agricultural and plant components. The plant data have received the greater amount of attention because more accurate estimates are possible, but even here there are enormous disparities between reports, and nearly all are based on engineering estimates of hypothetical distilleries rather than on actual plant operation data. Large discrepancies between estimates and actual results can be expected. Ethanol has a total energy content of 5553 kcal per litre. Traditional ethanol distilleries were estimated to require over 4000 kcal per litre for distillation only, so would confidently be predicted to have a massive negative energy balance, if fuel had to be brought in for distillation. Modern fuel ethanol plants have shaken off the traditions of the potable alcohol industry and have reduced distillation energy consumption by a number of measures: fermenting to higher ethanol concentrations, using stills based on designs of the petrochemical industry, more effective heat insulation, recovery of low grade heat as low pressure stream, and recovery of heat from stillage and transfer to incoming cold mash via heat exchangers. It is difficult to say with certainty what effect these modifications have had, and it almost certainly varies from plant to plant. Estimates are put as low as $0.95\text{--}1.92$ kg steam l^{-1} 99.5% ethanol (Keim, 1983), which, approximating steam at 800 kcal kg^{-1}, is 750–1500 kcal l^{-1}. Even then distillation is still the largest single consumer of process energy. An estimate for a modern maize-based ethanol plant (Coote, 1983) gives the following breakdown:

	kcal/litre
Maize cooking	432
Distillation	1630
Spent grain processing	894
All other parts of the process	63
Fermentation system including yeast propagation	12
Total	3031

The energy input for agricultural production of grains or other crops shows great variations according to estimate and location. They depend on factors such as the degree of mechanization, use of fertilizers, irrigation and so on. It is important to note here that use of marginal land almost invariably incurs a disproportionate increase in fertilizers to improve yields. Some idea of the details and assumptions involved in calculation and the variation according to location can be gathered from calculations on the economics and energy balance of ethanol from beet and cane sugar (Mouris, 1984). These estimates include detailed breakdowns of crop production, transportation, fermentation and distillation, plus considerations of wastes and by-products. Most authorities do not include the most comprehensive estimates of energies expended in constructing plant, heating and lighting of offices, transport of personnel to work and so on.

In general it can be concluded that ethanol from sugar cane has a clear advantage because of the distillation energy obtained from burning bagasse. Most authorities accept that it has a clear positive energy balance, though this may be dented by mechanical cutting and intensive irrigation. In Brazil it has been estimated the energy produced per hectare is 26 500 Mcal and energy consumed 18 500 Mcal giving a positive balance of 8000 Mcal or 30% of energy generated. Sugar beet, it is claimed, could have a positive balance, but it is marginal (Mouris, 1984). Molasses, because it is a by-product, can only be costed in energy terms for its transport. If efficient stills are used ethanol from molasses could have a positive balance. Grain grown for ethanol production is much more marginal. Distillation must be efficient, and transport costs low, and even then it is greatly dependent on the agricultural methods used – particularly with respect to fertilizers and mechanization.

The energy balance for ethanol from cellulose is almost certainly negative at present because of the energy intensive pre-treatment procedures required.

In the broader context of economic welfare the significance of negative energy balances or of the marginally positive balances which characterize most fermentation ethanol processes is difficult to appraise. If the negative balance relates to one liquid fuel *versus* another it is definitely disadvantageous. This was the case in the early days of corn ethanol plants in the United States where fuel oil was supplying energy for distillation. Another to be borne in mind here is that diesel grade fuels, often similar to bunker fuel oils, will give 50% more mileage than alcohol–gasoline blends. When other fossil fuels are employed, for example coal, it becomes an issue of generating liquid fuel or a more convenient energy form from a less convenient one. Factors such as whether the coal is imported, or what yields could be obtained from liquefaction, must also be considered. On this topic it is worth remembering that when coal or oil is used to generate electricity approximately 70% of the energy is lost in order to obtain a more convenient energy form. For example 1 te of oil, total energy content 44000 MJ, yields an average 4000 kWh or 14000 MJ electricity in a modern power station.

In microeconomic terms the energy balance is not important to an individual producer, providing he can produce ethanol at a profit using other fuels. In macroeconomic terms to a government, particularly one subsidizing an ethanol programme, the size and nature of the fuel balance is extremely significant. A nation with large, easily accessible reserves of coal, but a shortage of imported liquid fuels may consider ethanol fermentation worthwhile. Nations which either must import coal or effectively use coal and oil interchangeably, or which have limited coal reserves, would be ill-advised to support ethanol programmes at present energy balances. Smaller or older, less energy efficient plants should also not be supported.

Ethanol fermentation in the United States

In the aftermath of the 1973 oil crisis, fermentation ethanol received little attention from the United States government compared to projects for the liquefaction and gasification of coal, or even solar, wind or tidal energy, or solid waste combustion. Federal money allocated for ethanol production was primarily used for cellulose degradation projects. The agriculture lobby, faced with enormous surpluses and low prices for products, especially grains, launched a campaign to convert excess grain to liquid fuels. In response, in 1977 the US Government passed legislation exempting gasohol from the $0.04 per gallon federal excise tax on gasoline. Gasohol was defined as a mixture of at least 10% ethanol and 90% gasoline. Because each gallon of ethanol can be used to produce 10 gallons of gasohol the $0.04 per gallon exemption for gasohol translates into a $0.40 per gallon (or $16.80 per barrel) ethanol tax benefit. The ethanol must be generated biologically, but not necessarily from grain. In essence this benefit allows fermentation ethanol to be priced $0.40 per gallon above the price of high octane fuel and/or such additives as toluene, ethylene and benzene which are also used as octane boosters. The tax relief, originally scheduled to expire in 1984, was extended to 1992 in 1979, and was raised to $0.05 per gallon gasohol as federal gasoline taxes were raised in December 1982.

In addition, in response to the rise in oil prices in 1979–80 the US government introduced various incentives aimed at increasing production of fermentation ethanol. In particular a double investment tax credit of 20% was enacted for investment in ethanol production facilities. At state level too, various states, particularly in corn-growing areas, have exempted gasohol from state excise taxes (normally at $0.05

Table 8.9. *US corn production and carry over*
(figures in millions)

| Year | Production | | Year end carry over | |
	Bushels	Tonnes	Bushels	Tonnes
1967	4860	123.4		
1972	5573	141.6		
1977	6505	165.2	1111	28.2
1978	7268	184.6	1304	33.1
1979	7939	201.6	1618	41.1
1980	6644	168.8	1034	26.3
1981	8201	208.3	2076	52.8
1982	8315	211.2	2791	70.9
1983	(4500)[a]	114.3		

[a] Estimate.
From: Keim (1983).

per gallon) provided the raw material and the ethanol are produced
within the state. These states include Colorado, Kansas, Iowa, Illinois,
Missouri, Michigan and Wisconsin. Ethanol generated from corn and
whey in dairying states such as Wisconsin now attracts in effect a $1.00
per gallon tax exemption or $42 per barrel. When it is recalled that
present crude oil prices are $28–30 per barrel this is a remarkable
economic incentive. It should be noted that ethanol produced from
molasses, which is mostly imported, is eligible for the federal, but not
the state, tax exemptions.

To understand the reasons behind these moves it is worth
considering, in addition to oil price rises and strategic thinking, the US
production of corn (Table 8.9). Corn production in 1982 was 70%
higher than in 1967 and the carry over (i.e. surplus stored) from one
year to the next has more than doubled between 1977 and 1982 to over
70 million te. The 1983 figures are lower, due to a severe drought, and
will no doubt reduce the carry over figures. In the mid-1970s the
United States was forced to sell large quantities of grain to the Soviet
Union, in some cases because it simply could not be stored. In addition
the corn price, when corrected for inflation, has fallen considerably
since 1950, despite fluctuations (Fig. 4.1). This represents a big fall in
price when corrected for inflation. Only approximately 10% of the
entire US corn crop is used for human food or making human food
products such as starches, sweeteners, corn meal, beer and other
alcoholic beverages. The protein portion of processed grain and the
remainder of the crop is used to feed animals (60%) or exported (30%),
where in turn it is mostly used for animal feed. The subsidies were

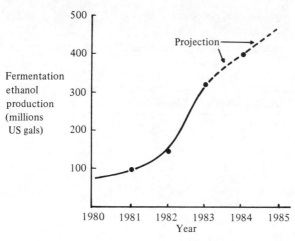

Fig. 8.6. Production of fermentation ethanol for fuel in the United States.

originally calculated in light of this grain surplus, since some US farmers have been paid not to produce corn at the rate of $1.00 per bushel that the land could have produced. Since one bushel of corn will yield approximately 2.5 gallons of ethanol it was felt that rather than pay a non-production subsidy of $1.00 per bushel, a $0.40 per gallon ethanol tax exemption would be preferable and have an equivalent cost.

In view of these measures it is hardly surprising that US production of fuel ethanol has risen dramatically (Fig. 8.6). In 1983 US fermentation ethanol production exceeded 300 million gallons (or 900 000 te) which was greater than synthetic ethanol production (at 700 000 te) for the first time since the late 1940s. Production capacity is now over 400 million gallons and predicted to rise to 500 million by 1985. Gasohol now accounts for 3–4% of total US gasoline sales, which means that ethanol even at these levels is only contributing 0.3 to 0.4% of total gasoline consumption, but this requires only 120 million bushels of corn of a total annual production of 8000 million. If all US corn production were converted to ethanol it could provide *c*. 20% of US gasoline needs.

The original US government policy was to encourage development of small scale (10 000 to 1 million gallons per year) plants, many of them on farms, in addition to the medium and large scale operations developed by wet and dry corn milling firms. The Department of Energy forecasts were for 300–400 small scale plants in 1982 rising to as many as 2500–3000 by the mid-1980s. In fact these developments have failed to materialize. Many plants either failed to start up or started but closed down, so that by 1982 there were only 50–100 small plants producing some 15–20 million gallons of fermentation ethanol per year,

Table 8.10. *Size of fuel ethanol plants in the USA (1982)*

Plant size (gal. per year)	Number of operating plants	Total capacity (million gal per year)
100 000–499 999	14	3.66
500 000–999 999	19	13.19
1 000 000–2 499 999	14	22.30
2 500 000–4 999 999	9	29.98
5 000 000–9 999 999	3	21.00
10 000 000–49 999 999	1	40.00
50 000 000+	5	340.00
Total	65	470.13

From: J. Haggin & J. H. Krieger (1983). Biomass engineering becoming more important in US energy mix. *Chemical and Engineering News*, **61 (11)**, 28–30.

with little increase in prospect. Lyons (1983) has attributed these failures to three factors:

(1)   Economies in capital cost were made by purchasing substandard equipment such as mild rather than stainless steel. Corrosion problems were encountered within months of start-up. Plants were poorly designed.

(2)   The small producers often found markets were slow to develop, they were unable to cope with cash flow problems in the first six months and filed for bankruptcy. Suppliers then considered this a high risk area.

(3)   Operators had no experience in the fermentation industry. Insufficient cleaning led to infection. High temperatures inactivated the yeast. Essential concepts such as pH were not understood by operators.

In contrast, large manufacturers, particularly corn wet millers, have expanded rapidly. Over 90% of production capacity is owned by nine major producers. One producer, Archer Daniels Midland, owns 220 million gallons capacity (i.e. 50% of the total) at its three locations, yet it only takes less than 20% of its corn starch to ethanol, the rest being used as industrial starch, dextrins, glucose and fructose syrups. The detailed economics of corn wet milling plants are discussed in Chapter 9. A distribution of size of US fuel ethanol plants is shown in Table 8.10. The six largest plants with capacity over 10 million gallons per year provide over 80% of the total capacity. These large plants, by benefiting from economies of scale, can be built for $1.75 per annual

Table 8.11. *US ethanol plants – economy of scale (1982)*

Plant size (millions of gal/yr)	Investment		Operating costs (millions $ per year)			Operating costs ($/gal)		
	(×$10⁶)	($/gal)	Fixed[a]	Payroll[b]	Total	Fixed	Payroll	Total
10	40	4.00	6.4	1.8	8.2	0.64	0.18	0.82
50	110	2.20	17.6	3.7	21.3	0.35	0.07	0.42
100	175	1.75	28.0	4.8	32.8	0.28	0.05	0.33

[a] As percentage of investment. Depreciation 10%, maintenance 4%, property taxes and insurance 2%.
[b] All plant personnel.
From: C. R. Keim (1983). From public information on Federal Loan Guarantees and from Keim's personal experience with various projects.

gallon of production, which translates to fixed and labour costs of $0.33 per gallon, as opposed to $0.82 for a 10 million gallon per year plant which is still mid-sized (Table 8.11). The same author concludes that only large wet milling plants can cover basic factory costs, overheads and selling expenses and yield a return on investment sufficiently above current interest rates to reward the risks taken (Keim, 1983).

Ethanol fermentation in Brazil

Brazil is the world's leading manufacturer of fermentation ethanol, producing 5.8 billion litres in 1982/83 (4.5 million te). Until 1975 the alcohol industry was an offshoot of the sugar industry, but even so production totalled some 600 million litres in 1975, some of which was used in 5 or 10% petroleum blends. In 1975 the Brazilian government instituted the National Alcohol Programme (PNA) to promote ethanol production from renewable resources and to displace imported petroleum, both as a fuel and a chemical feedstock. There were several objectives in this programme, which was in part in response to the oil price rises of two years earlier. The first was to reduce foreign exchange losses in an effort to combat the country's then large balance of payments deficit. Brazil also has huge foreign debts, which in 1982 stood at $65 billion and in 1984 have been estimated as high as $96 billion. The cost of servicing the debts in 1982 was $19 billion, $10 billion in interest and $9 billion in amortization. Subsequent rises in interest rates meant that the country could only maintain interest payments and then only by the implementation of austerity programmes with severe social consequences. Oil consumption in 1980 was 600000 bbl per day (30 million te per year) and accounted for 46% of the value of all imports. This had risen from 9% in 1970 to 25% in

1975. The country was importing 45% of its energy needs in 1979/80 and was crippled by the oil price rises.

Brazil is a net food exporter with a huge land area (2.5 times that of Western Europe), only 4.2% of which is used for arable farming. It has a rapidly increasing population and needs to create 1.2–1.3 million new jobs each year. The current production of ethanol employs a direct labour force of over 400000, of which 300000 are in farming and 90000 in the sugar factories and distilleries. It has been claimed that almost a million people may have jobs indirectly as a consequence of the programme. The alcohol programme also supports the sugar cane industry, which has suffered badly through periods of sugar oversupply on world markets.

The government support for ethanol production from sugar cane must be understood against this background. Support is implemented by fixing both the price of imported petroleum and the local price of sugar so that both are favourable to ethanol production. The government can control this production by raising or lowering the guaranteed price levels, i.e. the petroleum price. It is estimated that the production cost of ethanol is some 30% higher than imported petroleum before taxation is applied, i.e. it is approximately $40 per barrel. Before the 1979 price rises it was over 100% higher. The price of ethanol as a chemical feedstock is fixed by government subsidy at 35% the price of ethylene on a weight basis. Incentives are also offered to owners of cars which run on neat ethanol (95%). The owners pay lower taxes and can obtain better finance terms for new vehicle purchase. Much of the capital for distilleries is raised privately and they are in private ownership, but subsidized loans are available for agricultural development and new distilleries.

After 8–9 years of operation the Brazilian ethanol programme is still viable and still roughly on course, despite some shortfalls on targets which were always optimistic. Production has risen steadily to 16 times its level in 1975/76 (Fig. 8.7). All Brazilian cars run on ethanol fuel, 0.75 million on neat ethanol (95%) and the remainder on blends. The ethanol programme is estimated to have cumulative savings in foreign exchange approaching $5 billion since 1977 and the rate is now put at $1.3 billion annually. It can be argued to have made some contribution to Brazil's transformation of its once large balance of payments deficit to a small surplus by mid-1984, although overall petroleum consumption has carried on rising.

Sugar cane juice and molasses are the predominant feedstocks. Of the 380 authorized distilleries in 1982 only one was operating on cassava (manioc), although this crop is favoured by many for longer term development because of its growth on marginal land without extra fertilizer application. The distilleries are relatively small, with an

Fig. 8.7. Production of fermentation ethanol in Brazil. Productory figures from Copersucar (The Sugar and Alcohol Co-operative of Sao Paolo State) Annual Report 1984.

average output of 15 million litres (*c.* 11 000 te) ethanol per year. The limited harvest period means that daily capacities are higher, but over 50% have a capacity below 120 000 l per day and only a few above 200 000 l per day. They are usually annexed directly to a sugar cane processing factory and use molasses, or serve several surrounding cane factories using cane juice and molasses. This proximity is important in view of the high transportation cost of bagasse for distillery fuel and the rapid deterioration of sugar cane and cane juice.

The fermentation step has changed little in 40 years. It is conducted batchwise in open-topped 100–200 m³ vats. Distillation columns are made to a standard design with sections that can be bolted on and strapped to remain in position. Individual sections can be replaced if corroded. The hardware is nearly all fabricated in Brazil, it is simple, easy to repair and extensive use is made of low cost carbon steel. The fermenters are often coated with epoxy resin. Corrosion damage in fermenters must be repaired every 3–5 years and the units replaced every 8–10 years. The plants have very little instrumentation, other than temperature and steam pressure gauges, they have no feedback control and very definitely no computers. They are operated in a batch mode, and steam is generated by burning bagasse which gives them a positive energy balance; they only have evaporative cooling and no mechanical agitators. The alcohol programme is also labour intensive, both in distilleries and in crop production.

Table 8.12. *A comparison of ethanol fermentation in Brazil and the United States*

Item	Brazil	United States
Raw material	Sugar cane/molasses	Maize
Plant size	Small	Large
Costs	Labour intensive	Capital intensive
Construction	Carbon steel	Stainless steel
Mode	Batch	Continuous, fed-batch or cascade
Control	Manual	Computerized feedback
Energy	Not critical, abundance of bagasse	Conservation essential
Distillation	Simple, basic	Sophisticated, petrochemical technology
By-products	Virtually zero	Only operate as part of an integrated wet (or sometimes dry) milling process

By-product credits are small or non-existent. Raw stillage is used as a fertilizer for the mineral content. Waste disposal is a major problem since each litre of ethanol is accompanied by 12–14 litres of high BOD waste. The total effluent has been calculated to be equivalent to the sewage of a population of 160 million l (1.5 times the present Brazilian population).

Technologies compared

The strategies of the two major fermentation ethanol producing nations make an interesting comparison (Table 8.12). Both are underwritten by government and both use *Saccharomyces cerevisiae*, but there the similarities end. In Brazil the use of sugar cane has dictated small plants adjacent to the plantations. They are simple, robust in construction, easy to repair and easy to operate (Guidoboni, 1984). They are labour intensive, but this is an advantage to the country at present. In the United States small scale plants have lost out almost completely to large ones. Six plants provide over 80% of capacity and one manufacturer is responsible for over 50% of production.

The use of corn as a substrate permits transport storage and year-round operation. The American infrastructure, particularly the railroad network, permits the moving of bulk quantities (1000–200 te trainloads) of corn into plants and that of products away. Corn wet milling has proved to have superior economics to dry milling, largely through by-product utility (Chapter 9). Wet milling plants benefit from economies of scale and are generally limited by the grind capacity. The high degree of technology and skill in constructing and operating wet milling plants provides a formidable entry barrier to new competition and has led to concentration of production among a few firms. Ethanol plants also benefit from economies of scale and the US is a high labour cost country. Hence large sophisticated plants with a high degree of automation and feedback control have succeeded. They often operate in fed-batch or cascade modes. They are invariably constructed from high grade stainless steel. Distillation practices have been adapted from the petrochemical industry because of the marginal energy balance and the importance of fuel economy. Stills have high numbers of theoretical plates, energy recovery through heat exchangers, and some have mechanical vapour recompression. They are tall (over 100 ft), free standing structures in complete contrast to the Brazilian versions. They have high capital costs, but low operating costs, particularly in relation to energy. Capital costs have to a certain extent been defrayed by government tax concessions.

Despite these differences costs of ethanol production in the two systems are probably relatively similar. There are almost as many different detailed costings as there are authorities on the subject. The major differences between the Brazilian and US costings are the values of by-product credits and higher capital costs in the US (Tables 5.5 and 5.6). It is very difficult to put a value on the cost of producing cane juice, because this is dependent on its alternative use in the production of crystalline sugar. Since this is so cheap at present, sugar factories in developing nations can only barely cover marginal costs, hence opportunity costs are low. The situation may change in a few years. It may be more attractive to produce crystalline raw sugar. Similarly the economics of ethanol in the US plants cannot be dissociated from the alternatives of directing product streams into glucose or high fructose syrups or from the market value of by-products such as gluten in animal feed.

Other countries

Developments in biological fuels have not been confined to Brazil or the United States although these two nations have taken the lead. All non-petroleum exporting nations have adopted energy policies, but adapted to individual conditions and with varying degrees of efficacy. Raw materials or national assets are obviously the prime consideration. Many nations cannot consider biological feedstocks, unless major improvements are made in cellulose and lignocellulose degradation. They have inadequate land to produce cereal crops or are net importers of food.

India is the world's third largest producer of fermentation ethanol,

having approximately 100 distilleries with an installed capacity of 600
million litres (*c.* 470000 te) and production around 500 million litres *per
annum*. India is a country with over 600 million inhabitants, and it must
use all its cereal crops for human food, but it does produce over 2
million te of sugar cane molasses p.a. which cannot be used for human
food. In order to reduce petroleum imports, this molasses is directed
into fermentation ethanol production by legislation which prevents
export and fixes a low transfer price (under $10 per te) between sugar
refiners and fermentation operations. The average yield of ethanol per
te of molasses (48% fermentable sugars) is approximately 225 litres.
The ethanol is not used for fuel; two-thirds is for industrial use and
one-third is potable. The industrial usage is for ethylene derivatives
(PVC and polyethylene), organic chemicals and synthetic rubber. Even
with the low molasses prices synthesis costs are still higher than from
petroleum and increasing self-sufficiency in oil may induce the country
to re-examine its ethanol policy.

The European Economic Community is a net importer of cereals
despite high domestic support prices of most agricultural products.
Over 20 million te of feed grains, mostly maize, are presently imported.
France and Germany both produce ethanol from sugar beet molasses.
France has some 19 distilleries which produce 160 million litres
(125000 te) of ethanol entirely from sugar beet. They employ
approximately 1000 people. The state 'Service des Alcools' buys 80%
of production at a fixed price based on the guaranteed price for sugar
beet plus a manufacturing margin. Certain industrial end users are
obliged to purchase fermentation and not synthetic alcohol. Japan, like
the EEC, is a net importer of agricultural products. It formerly
produced 40% of its ethanol consumption by fermentation from
imported molasses, but is increasingly importing crude fermentation
ethanol from sugar cane growing countries, mostly Thailand and the
Philippines. Similarly the Sovient Union has a chronic grains deficit and
rising animal feed demand. COMECON countries import over 20
million te of grain annually. The Soviet Union has a huge land area,
but a poor climate with low overall temperatures. It is also at present
the world's largest oil producer (616 million te in 1983 *versus* 430
million te from the USA and 260 from Saudi Arabia). There is
therefore little incentive to produce fermentation ethanol. Rather the
country is building single cell protein plants using petroleum feedstocks.

Other nations with spare agricultural capacity are developing ethanol
programmes, though on a much more restricted and trial basis. Sweden
is constructing a plant to operate on wheat starch with animal feed
by-products. The Philippines are embarking on a programme to
produce fuel ethanol from sugar cane. Thailand is planning to develop
a cassava-based programme. Similarly Australia, with huge resources of

marginal land, but small oil reserves, is interested in cassava plus sugar cane (grown in Queensland) to produce ethanol. With abundant and cheap coal Australia's concern is for liquid fuels but new oil discoveries may change this.

Improvements to the fermentation ethanol process

The production of industrial or fuel ethanol by fermentation is at best a marginal process even at present oil prices. An enormous amount of research world-wide is being directed to improvements in the process. Costs are so dominated by raw material requirements that refinements to other parts of the process are only likely to result in savings to the order of a few cents per gallon. Even so, in the giant plants operated in the US, savings, even of this order, can translate to millions of dollars for the operators.

Unless cereal production costs drop dramatically the best prospects for lower cost raw materials lie in cellulose or lignocellulose. These are at present limited by high hydrolysis costs, either enzymic or in pre-treatment (see Chapter 9). There is also a need for industrial organisms, i.e. ethanol-tolerant, high productivity strains, to convert 5-carbon sugars to ethanol.

Aside from substrate the greatest need is for higher ethanol concentrations and shorter fermentation times. In the absence of better organisms these goals can be achieved by continuous culture and improved reactor designs. Both topics and their limitations have been outlined in Chapter 5. Given the problems with continuous culture, some ethanol manufacturers have adopted two other approaches to achieve at least some of the potential economies. The cascade system (Fig. 8.8) can give many of the advantages of continuous operation; the tank volume is the same as in a batch system, but there is no down time for vessel emptying and cleaning. It has the disadvantages of requiring more pumps than a batch system and the first tank operating at a low ethanol concentration is liable to infection. If infection does occur it then cascades through the system. However this type of system is operated successfully by some US firms where infection is minimized by use of antibacterial agents. Normally up to five fermenters are used in series with final ethanol concentrations of 11–12% after a fermentation time of 18–24 h. The tower system counterbalances an upflow of mash in a tower, with the rate of fall of a flocculant yeast, yeast concentrations and residence times being set so that the mash is fully fermented by the time it reaches the top of the tower. This system has been used successfully in beer manufacture, where 3–4% ethanol can be made satisfactorily, but 10–12% mashes for distillation have not been easy to control. Fermentations were incomplete and yeasts were lost with mash at the top of the tower.

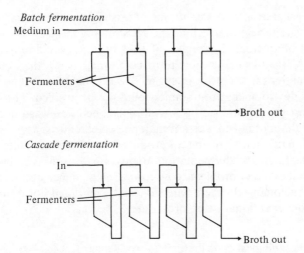

Fig. 8.8. Batch and cascade fermentations. From Coote (1983).

Continuous ethanol removal is another long sought after target, either by operation at higher temperatures and/or reduced pressures. Vacuum systems have high energy requirements for cooling and pumping (carbon dioxide recompression). Flash distillation has similar drawbacks. None is operated on a commercial scale at present. Membranes which are selectively permeable to ethanol have also been proposed, but are very much in an early experimental stage. Operation at high temperatures is dependent upon the selection of thermophilic ethanol-producing bacteria which, like yeasts, convert over 90% of substrate to ethanol. Thermophiles will have rapid metabolic rates and reduced contamination problems, but all found so far produce lower ethanol yields than yeast, and other by-products such as acetic and lactic acids, and they have low ethanol tolerance. The more limited goals of yeasts operating at elevated temperatures (40 °C) and with higher ethanol tolerance seem attainable. Thermotolerant yeasts could overcome cooling water requirements in tropical or subtropical locations, and the incorporation of unsaturated fatty acids into yeast membranes improves ethanol tolerance. Either advance will only have marginal effects on the economics of ethanol production.

In spite of the enormous progress in microbiology in recent decades, the large number of new organisms isolated and the advances in genetics, traditional yeasts still have no serious challengers. *Zymomonas* strains do have higher specific ethanol production rate, but lower ethanol tolerance. A recent study on the sensitivity to genetic engineering of the economics of the ethanol fermentation (Flannery & Steinschneider, 1983) concluded that the greatest potential for genetic engineering improvements lay in providing less expensive raw materials through the generation of better strains of existing crop plants, new agricultural varieties and the cellulose-degrading organisms. Achievement of an ethanol tolerance of 12% or 20% would only give 0.6% or 1.3% savings in manufacturing cost, which is less than the normal fluctuations in the price of sugar. It was concluded that such a research and development programme would not be justified. The maximum cost saving attained by thermophilic operation was put at 5.1%, again thought not to warrant major research efforts.

In general, particularly to ethanol producers in developing nations, improvements to ethanol fermentation involve greater sophistication, higher capital costs and problems concerning maintenance of equipment. The financial benefits are sufficiently marginal not to justify departure from traditional batch procedures.

Biotechnological contributions to world energy needs – an overall appraisal

Biotechnological processes for energy generation are only operated on any scale with political intervention in one form or another. Ethanol programmes are operated by Brazil, the United States and India. Acetone–butanol survives in South Africa, a country with abundant coal and molasses but with oil problems. Without subsidy, fermentation ethanol has not been competitive with petroleum products for fuel or chemical feedstock use for at least 40 years. Even today petroleum prices will have to increase to the $50–60 per bbl range and perhaps higher, and all costs in ethanol production, including raw materials, will have to remain constant for the two to compete in an open market. If this were to happen – and in the light of events in the 1970s, it could – many other fuels would come into economic range, notably liquefaction or gasification of coal and extraction of oil from tar sands or shale. The earlier price rises have already displaced much oil used in electricity generation by coal and nuclear fission. Oil prices are soft at present because of economic recession, but they may rise if the world economy recovers. Oil prices have already peaked at $35 per barrel before falling to present levels. On the other hand it is in the interest of oil exporters to drive the price up to (but not above) the cost of alternatives. If the price of oil rises in real terms faster than the major customers can absorb it, the resulting recession will create an oversupply which will prevent further price rises. As many alternatives appear to be cheaper than fermentation ethanol, unless radical process improvements are made the immediate future must lie in subsidy or higher value applications.

It is very difficult to appraise the impact of the government subsidies on general welfare. They are similar to protectionist policies in agriculture or manufactured goods. The home market is guaranteed and employment is created, but the consumer pays higher prices, either directly or through taxation. Some wealth is recovered directly in wages and taxes paid by the subsidized manufacturer, but the big question mark is the opportunity cost to the community. Both the Brazilian and American programmes are classic illustrations of the dilemma. The Brazilians have invested $5–6 billion in the alcohol programme. They save 20% of their petroleum imports, but consumption has still risen. Could the money, if invested elsewhere, have achieved more for the economy, say in a direct foreign debt repayment or other industrial or agricultural projects, particularly aimed at limiting malnutrition? Brazil must still import 6 million te grain despite having net agricultural exports of over $7 billion annually, although it can also be argued that there is no conflict over the land used for sugar cane. The Brazilian programme is also criticized on the basis that the subsidies have actually increased the number of cars and that the urban car-owning better-off are being underwritten at the expense of the rural poor. Could petroleum imports be cut by 10% directly by higher taxes or greater conservation, say via small or diesel cars? There is also now an urgent need for land reform, since the programme has contributed to increasing concentration of ownership and has channelled public aid into private ventures. On the other hand the ethanol programme has certainly been beneficial to the Brazilian sugar industry, giving it flexibility in diverting the crop away from crystalline sugar exports at a time of record surpluses and low market prices.

The American subsidy in the form of gasoline tax exemptions of up to $1.00 per gallon of ethanol amount to a subsidy on corn of $2.50 per bushel, which is approaching its market price. The displacement of gasoline in US markets is only approaching 0.4%. The cost to a wealthy nation such as the US is very small, while it stimulates new technology and helps reduce an embarrassing grain surplus. However it does not go far in solving energy problems. Similarly biogas generation can only have a very minor impact at present, but legislation permitting sale of electricity to the grid (Chapter 10), while only an irritant to utilities, could stimulate development of the field.

If biotechnological fuels are to be taken seriously and provide a significant contribution to world energy needs, then the problems of scale need examination. Fuel markets are hundreds of times greater than the traditional ethanol markets supplied to date. To make a material contribution, fermentation ethanol must be produced in large volume, and efficient plants assured of a continuous supply of low cost raw materials. Only nations with a net agricultural surplus, but importing energy (Table 8.13), would appear to benefit at present. The land required for cereal crops at their current yields provides another

Table 8.13. *Nations with net agricultural surplus but importing energy*

Dominican Republic	Philippines	USA
Cuba	Argentina	France
Thailand	South Africa	
Brazil	Kenya	

constraint. World oil consumption of 2800 million tonnes p.a., if replaced by an equivalent volume of ethanol produced from maize, would require approximately 10 million square miles of maize. The land areas of the Soviet Union, USA and China are respectively 6.7, 3.6 and 3.0 million square miles. Similar calculations can be made for individual countries: 1982 US maize production of 8315 million bushels would only supply 20% of its gasoline needs, while West Germany would require all its land suited for sugar beet production to produce 10% of its gasoline needs (Esser & Schmidt, 1982).

The third major constraint is the impact fuel ethanol requirements would have on agricultural commodity prices. The economics of corn wet milling are dependent on by-product credits and thereby on the maintenance of animal feed prices. There is a high correlation between these prices and the price of corn, so the credits have remained constant as a factor of the corn price (Keim, 1983). It seems unlikely however that demand for all the corn by-products could keep pace with ethanol demand as a major fuel. There is also the effect on other crops. Displacement of other crops will increase their price and lead to a competitive price spiral won by the use with the most inelastic demand.

These and other considerations on cereal crops, not least fuel/food arguments, lead to the conclusion that cellulosic and lignocellulosic biomass look the best bets for long term feedstocks for biological fuels (Table 4.7). It must also be remembered that crops are only as renewable as the plant nutrients in the soil. Improvements in the energetics of fertilizers, particularly nitrogen, are essential if even a moderately favourable energy balance is to be achieved. Otherwise large quantities of fossil fuel are going to be required to sustain a biological fuels regime.

In the short and medium term ethanol appears to be the only biotechnological product with any significance as a fuel. The relative changes in oil and agricultural commodity prices have meant that its price relative to oil is lower than in the past, but it is still not competitive without massive political intervention. Constraints on its volume of production by present technology from cereal crops mean that its uses are realistically restricted to an octane improver, in which role it could command a premium price in replacing the toxic tetraethyl lead. As a chemical feedstock ethanol is likely to remain at a severe disadvantage to petroleum ethylene in most syntheses, save perhaps oxidations to aldehydes or acids. The present poor economics of the acetone–butanol fermentation mean that it cannot compete without at least a doubling in oil prices or drastic improvements in productivity and product concentration.

In the long term, as with other areas of biotechnology, potentially the sky is the limit; this is particularly so with reference to photosynthesis, but these areas are currently beyond the scope of any detailed economic evaluation.

Further reading

The Alcohol Economy: Fuel Ethanol and the Brazilian Experience. (1983). H. Rothman, R. Greenshields & F. R. Calle. London: Frances Pinter.

Gasohol, A Step to Energy Independence. (1981). T. P. Lyons. Lexington, Kentucky: Alltech Technical Publications.

Organic Chemicals from Biomass. (1981). I. S. Goldstein. Boca Raton, Florida: CRC Press Inc.

Fuel from Farms – A Guide to Small Scale Ethanol Production. (1980). Solar Energy Research Institute, US Dept of Energy, Oak Ridge, Tennessee.

BP Statistical Review of World Energy. Published annually by the British Petroleum Company p.l.c., Britannic House, London EC2Y 9BU.

The Political Economy of Oil. (1980). F. E. Banks. Lexington Ma, USA: Lexington Books.

Oil and World Power, 5th edn. (1979). P. R. Odell. Harmondsworth, UK: Penguin Books.

World Energy Supply, Resources, Technologies, Perspectives. (1982). Grathwohl. Berlin: Walter de Gruyter.

The Chemical Industry. (1981). Ed. D. H. Sharp & T. F. West. Chichester: Society of Chemical Industry, Ellis Horwood Ltd.

9   The American corn wet milling industry

Background

In Chapters 5 and 8 it was shown that ethanol production costs are dominated by the cost of raw materials, but when maize is used these costs are effectively reduced by the sale of by-products. The by-product credits are highest in the wet milling process (Keim, 1983). In Chapter 7 the use of enzymes to hydrolyse corn starch to glucose syrups and their subsequent isomerization to high fructose syrups was described. High fructose syrups with production in the 3.5–4 million te range have displaced invert syrups as nutritive sweeteners in large sections of the US soft drinks market. The increase in sales volume of these products has led to corn wet milling becoming one of the fastest growth industries in the United States since the early 1970s. This is significant in that corn wet milling represents the first example of a modern, integrated, biotechnologically-based industry producing a range of high volume, low cost commodities from an agricultural feedstock.

The growth of this industry is a product of political, economic and agricultural forces coupled with advances in enabling technology. The United States has had a policy of encouraging maize production, which coupled with improved agriculture has led to large surpluses. As a consequence grain prices, when corrected for inflation, have fallen considerably, particularly in relation to oil (Fig. 4.1). The surpluses have also on occasion proved to be an embarrassment to the government. Cane and beet sugar are produced in the United States, but the climate is not ideal. Domestic sugar production is protected by a price support programme which translates to an effective price of $560 per ton for major industrial users. This also protects indigenously produced nutritive sweeteners from world market forces. Ethanol production is supported by the federal and state gasoline tax credits on gasohol blends (Chapter 8).

Milling processes

Maize is converted into food products by two methods, dry and wet milling. In addition there is the whole grain process traditionally used in whisky distilleries. In the whole grain process the kernels are crushed

214

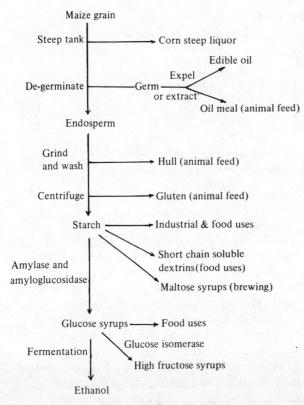

Fig. 9.1. Corn wet milling.

in roller mills or pulverized in hammer mills and mixed with water to make a slurry before cooking and hydrolysis. The unfermented materials are recovered and dried together as distiller's dried grains and solubles. This product contains at least 27% protein plus appreciable amounts of oil and fat and is a valuable animal feed. In most modern dry milling processes the corn is tempered by being soaked in water for a short time where the grain is broken loose and separated. The grain is used to make corn oil and the residual cake after oil removal sold for animal feed. The endosperm and fibrous hull are separated and the endosperm ground; the ground endosperm may then be used for making coarsely ground meal (grits). Endosperm is composed primarily of starch plus protein and among other uses is employed as an ingredient of beer, where the starch is saccharified and fermented along with barley malt. Endosperm is rarely used for fuel alcohol production, because the ethanol yield is lower than in the whole grain process and any added by-product values are insufficient to compensate for higher investment and operating costs.

Table 9.1. Wet milling yields

Products	Yields dry solids (lb) per bushel[a]
Feed: 21% protein	9.2
Gluten meal: 60% protein	2.7
Germ[b]	3.5
Starch	31.5

[a] A bushel of corn at 15.5% moisture weighs 56.0 lb (47.32 lb dry solids).
[b] 45–55% oil.
From: T. P. Lyons (1983). Ethanol production in developed countries. *Process Biochemistry*, **18**, 18–25.

Corn wet milling is a complex process used for the manufacture of corn starch and the various sweeteners obtained from its hydrolysis (Fig. 9.1). The corn grains are initially soaked in hot water with sulphur dioxide present to loosen the components. The grain is ground quickly to release the germ, which is treated to obtain corn oil, either by expulsion or solvent extraction. Residual oil meal is used for animal feed. The endosperm is ground finely and washed to remove the hull (husk). The residue is centrifuged to separate starch and gluten. Gluten is used predominantly as an animal feed although some is incorporated in human food products. There are two grades, gluten feed (minimum of 21% protein), which includes the hulls and various recycled products, and gluten meal, a premium product containing 60% protein used particularly in poultry feed. Until recently the restricted applications for starch were a commercial limitation on the process. Starch is dried and sold for industrial uses such as printing (60%), textiles (10%) and food uses (30%). In many food uses it is hydrolysed to intermediate chain length dextrins. There are also modified starch products (from high amylose corn and waxy maize), but they represent less than 1% of wet milling production. The yields of products per bushel of corn are shown in Table 9.1.

The bulk of starch is now used for the production of maltose syrups, glucose and glucose syrups, high fructose syrups and ethanol. In addition carbon dioxide produced by fermentation is sometimes used in hydroponic plant culture. Wet milling is thus a multi-product process with the many inherent commercial values and drawbacks that this confers. No one product can be considered in isolation. Increased production of ethanol or high fructose syrups can only be achieved if increased volumes of gluten can be sold at certain price levels. On the other hand surplus capacity of glucose syrups can be diverted to either high fructose syrups or ethanol. Different plants often have different product mixes.

Table 9.2. *Estimated US theoretical high fructose syrup production capacities, 1983*

Company	42%	55%	Total	Share (%)	Number of plants
Archer Daniels Midland	614	863	1477	35	4
A. E. Staley	420	893	1313	32	4
Cargill	190	259	449	11	2
CPC International	280	113	393	9	3
American Maize	75	136	211	5	1
Hubinger	45	100	145	3	1
Amstar	48	80	128	3	1
Great Western	14	32	46	1	1
Total	1686	2476	4162	100	17

Capacities are expressed in 1000 te dry solids
Based on a personal communication from P. S. J. Cheetham

Plant costs

Wet milling plants are capital intensive; they benefit from economies of scale and they occupy huge land areas. The most expensive part of the plant is its grind capacity, so plants have tended to have more finishing than grind capacity to enable production flexibility. Ethanol fermentation has greatly improved the flexibility because demand for the other major products (corn syrups, glucose syrups and high fructose syrups) has cross elasticity. The pricing and margin trends do not differ significantly in individual markets. A plant to produce high fructose syrups has capital costs in the region of $100 million for a capacity of 30000 bushels per day (500000 te yr^{-1}) and $160 million for 60000 bushels per day. Ethanol fermentation facilities add extra costs. The average size of plants is in the 300000–500000 te range, although several are now larger. An approximate breakdown of the high fructose syrup market in the United States and the number of plants involved is shown in Table 9.2. The pattern is very similar to US ethanol plants (Table 8.10).

History of development

The commercial development of high fructose syrups followed the discovery of glucose (xylose) isomerases in the 1960s, since previously available alkaline hydrolysis produced undesirable by-products and colour formation. The first full scale plants making 42% fructose syrups came onstream in 1970 and the pioneering companies (A. E. Staley and Clinton Corn) were able to make a killing during the sucrose shortage in 1973–74 when sucrose prices soared (Fig. 4.5). Their success in this

Table 9.3. *Matrix of pre-tax profits (loss) per bushel of corn grind*

| Selling price | | Gross price per bushel of corn | | | | | |
| | | $2.50 | | $3.00 | | $3.50 | |
Ethanol[a]	55% HFCS[b]	Ethanol	HFCS	Ethanol	HFCS	Ethanol	HFCS
1.70	15.00	1.00	(0.25)	0.75	(0.50)	0.50	(0.75)
1.70	20.00	1.00	1.40	0.75	1.15	0.50	0.90
1.70	25.00	1.00	3.05	0.75	2.80	0.50	2.46

[a] Ethanol prices in $ per US gallon. Pre-tax profits per bushel are based on 2.5 gallons yield per bushel of corn.
[b] HFCS: high fructose corn syrups. Prices in $ per hundredweight (100 lb). Pre-tax profits per bushel are based on a yield of 33 lb per bushel.
From: Greditor (1983).

period encouraged a major construction boom in high fructose syrup plants by the major wet milling companies and others, which led to overcapacity by 1977 because of the restoration of sugar supplies and a precipitous decline in sugar prices. In 1977–79 plant utilization levels were as low as 50%, despite a rise in high fructose syrup sales from 942000 te in 1977 to 1543000 te in 1979. This was therefore a period of intense competition between the major manufacturers, with margins cut to the minimum. By 1980 there was a big upswing in profitability because of a rise in sugar prices, the introduction of 55% fructose syrups and the decision by soft drinks manufacturers, notably Coca Cola, to replace sugar usage with these 55% syrups (Greditor, 1983). The fall in world sugar prices in the 1980s has not had the same impact on corn wet milling companies, in part because of their ability to switch to ethanol production.

Ethanol prices rose to $1.85 per gallon in 1980, fell subsequently to $1.60, but then stabilized at around $1.70. This price was assumed in drawing up a matrix of comparative returns from ethanol and high fructose syrups at different corn and high fructose syrup prices (Table 9.3). These data refer to early 1983 when the price of 55% fructose syrups was below the equilibrium price of around $18.90 per cwt and hence ethanol was a more profitable product. Since then the US government has increased both the sugar support price (to $28 per cwt or $560 per ton) and increased the federal gasoline tax relief on gasohol by 1¢ per gallon (10¢ per gallon of ethanol). The equilibrium position on the matrix has therefore changed in favour of high fructose syrups as their price rise has been greater, but the table serves to illustrate the relationship between the two products. It shows the basis for manufacturers' decisions on where to direct production.

Although ethanol production capacity involves extra capital costs,

Table 9.4. *Food, seed and industrial use of corn, 1980/81–1983/84*

Products	1980/81	1981/82	1982/83[a]	1983/84[a]
HFCS	160[b]	185	215	240
Alcohol[c]	70	125	180	200–240
Glucose and dextrose	182	185	186	187
Dry milled for food and beer[d]	173	172	177	176
Starch	130	125	125	128
Seed	20	19	17	19
Total	735	811	900	950–990

[a] Forecast.
[b] Million bushels.
[c] Fuel, industrial and beverage alcohol.
[d] Corn meal, grits, flour, snacks and speciality food.
From: *Sugar and Sweetener Outlook* and *Situation*. US Dept. of Agriculture Economic Research Service, June 1983.

process improvements in recent years, notably in distillation, have reduced operating costs. There are also other benefits in ethanol production. It can be produced by the fermentation of waste and off-grade sweetener products and it can assist wet milling companies to overcome the seasonal fluctuations in demand for high fructose syrups which are associated with their use in soft drinks (Greditor, 1982). As a result ethanol production has risen rapidly, far outstripping the rise in high fructose syrup volume (Table 9.4). Starch and glucose syrups are now mature products with constant volumes of production. Corn products are also now being looked at as feedstocks for the chemical industry. For instance methyl glucoside, used in polymers in foams and paints, is now being manufactured by A. E. Staley as a first step in diversifying its starch stream into chemical manufacture.

Wet milling outside the United States

Japan is a net importer of both corn and sugar and has moved into high fructose syrups because, unlike sugar, they are not subject to import duty and taxes. Sucrose solutions are frequently blended with these syrups. Total sales of high fructose syrups were approximately 600 000 te (dry solids) in 1983, but wet milling is not used to produce ethanol in Japan.

European-grown corn is not suitable for wet milling, so the industry is based on imports from North America. The rapid expansion of wet milling capacity in Europe in the 1970s provoked protectionist action from the European Economic Community sugar beet producers, who already have a surplus production which is sold off cheaply on world markets. As a result the EEC introduced a quota system on the production of high fructose syrups in 1979 after a levy imposed in 1977 was shown to be illegal. This action has kept high fructose syrup production at under 200 000 te with capacity around 250 000 te. As in Japan there is no ethanol production.

Canada can use indigenous maize, but does not produce sugar and does not support sugar prices in the way that the United States and the EEC do. As a consequence high fructose syrups have had to compete with sugar on the open market. There are some wet milling plants, but they are far fewer than in the US and have much poorer profitability. Small plants have been constructed in Hungary, Yugoslavia and Pakistan but production levels are still very low.

Summary

The corn wet milling industry has achieved rapid physical volume growth in recent years. It is a very capital intensive industry with high costs associated with the early milling steps. As a consequence the pressure to maximize utilization rates is intense and leads to severe price competition as individual producers try to maximize their market share. The industry has suffered from a roller-coaster effect on profitability from expansion and subsequent overcapacity, which has been exacerbated by the cross elasticity of demand between its products and refined sugar. Ethanol production has aided the industry because it can generate incremental earnings independent of the sweetener market.

The wet milling industry has thus progressed to a range of products by making new additions as older products reached maturity. Ethanol and high fructose syrups produced by this industry are two of the major volume products of biotechnology. High fructose syrups are the largest product of immobilized enzyme technology. They have made major inroads into the carbohydrate sweeteners market in the USA and Japan and to a lesser extent in Canada and Europe. Ethanol now accounts for 0.3–0.4% of United States automobile fuel consumption.

The future of the industry is far from clear, however. It is in a stronger competitive position for ethanol fermentation than dry milling because of by-product values. The growth of high fructose syrups depends on the substitution rate in soft drinks, which is now beginning to approach saturation and sales are expected to plateau. Ethanol production can only continue as long as the federal and state gasoline tax exemptions are maintained. If they were to be stopped or reduced the resulting overcapacity would swamp the sweetener market and probably cause plant closures or even company failures. On the other hand increase in alcohol production will affect a complex system of economic relationships. There is the response to increased demand for the grain itself and the response to increased supplies of the by-products on the animal feed markets. Increased demand for grain may only have a limited effect on prices initially, but if a very large scale programme were introduced, say the 10% fuel substitution once planned, it would increase prices and make the ethanol programme expensive, if supportable at all. Large quantities of by-product feeds can be utilized at higher price levels if the export market continues. Any imposition of duties or glut of animal feed could force lower relative prices and again jeopardize the ethanol programme.

Similarly adoption of a high fructose syrup programme by other countries can only be at the expense of sucrose. The current spot price of world sugar is below even the most efficient producers' costs of production. They are operating on marginal costing criteria unless they have fixed price contracts. Under these circumstances high fructose syrups can only be profitable when protected by tariffs or government support programmes. Unless there is a surplus production of maize there seems little point in the substitution of one product by a virtually identical one. Other countries, notably some in Europe, are surplus producers of other grains such as wheat and barley, but wet milling processes or processes of equivalent efficiency are not available.

Further reading

An alcohol fuels programme; will it solve surplus grain production? B. Chattin & O. C. Doering, (1984), *Food Policy*, **9**, 35–43.

F. O. Licht's International Sweetener Reports, published annually in December by F. O. Licht, Ratzeburg, W. Germany.

Ethanol production by fermentation: an alternative liquid fuel. N. Kosaric, D. C. M. Ng, I. Russell & G. S. Stewart, (1980), *Adv. in Appl. Microbiol.*, **26**, 147–227.

10 Waste treatment

Introduction

In volumetric terms treatment of domestic waste water and sewerage is by far the largest biotechnological industry. In the UK, Dunnill (1981) estimates fermentation capacity for waste water to be $2\,800\,000$ m^3 compared to that of its nearest rivals beer at $128\,000$ m^3 and antibiotic production at $10\,500$ m^3. There is little to suggest that the picture is not similar in most industrial countries. In economic terms this system of sludge beds and aerobic filtration through beds containing microorganisms is operated by municipalities and charged to the inhabitants through the local taxation system. There are few if any by-product credits, other than occasional sales of treated sludge for compost or manure. Some authorities do operate anaerobic systems with methane collection, but they are very much in a minority. Historically, the system was forced upon the Victorian city-builders in order to combat the appalling epidemics of cholera and typhoid which swept through urban populations. Where alternatives existed, such as dumping at sea, these were frequently exploited, even though periodic outbreaks of diseases caused by enteric pathogens, notably poliomyelitis prior to the immunization programme in the 1950s, could be traced to bathing in polluted water. That this practice should continue today, even in resort towns, is one of the worst continuing public health scandals. It also serves to illustrate societies' attitudes to waste treatment and pollution. In general only when a problem reaches epidemic proportions are measures taken to combat it, and then only after long political campaigns. Similar developments in industrial waste treatment can be traced and they are only being combatted now following heightened public awareness of the toxicity of wastes in the last two decades. The essential need to treat wastes, not only for toxic chemical and organism removal but also to reduce biological and chemical oxygen demand (BOD and COD), is now recognized. Industries are being compelled to either cease dumping waste, or pay prohibitive penalties if they enter waterways or the municipal waste systems.

The system of penalties, generally based on COD, which is being adopted in many industrial countries is creating waste treatment

222

industries and is an opportunity for biotechnology. Some waste treatment plants which have been forced on operators have later been shown to be profitable in their own right, although the capital expenditure would never have been undertaken without legislative pressure.

Many categories of waste are amenable to biological treatment. Farm wastes, particularly from intensive livestock production, are now receiving much attention. Food processing plants frequently produce large volumes of high BOD wastes and are often targets for action by water authorities. As described in Chapters 5 and 6, biotechnological processes, particularly fermentation, also generate large volumes of high BOD wastes. Distilleries too pose a particular problem in that their effluent is concentrated with a high BOD and often high sulphate levels when molasses is the feedstock. Many paper and paper pulp wastes can be subjected to biological treatment and can sometimes generate useful products. Chemical plant effluent on the other hand tends to be more a case of clean-up and de-toxification. Household refuse presents a challenge and an opportunity for biotechnology because it is mostly biodegradable, produced in large volumes and a high proportion is cellulose.

Wastes may be viewed in three contexts. There is the pollution or hazard value, the potential for recycling valuable raw materials such as metals or possibly nitrogen and minerals for fertilizers, and the possibility of manufacturing products such as biomass, chemical feedstocks or fuels. Many approaches to waste treatment are being tested. Some, such as compression, incineration or physico-chemical recycling are beyond the scope of this review, but several biotechnological processes are important and may be subjected to economic analysis. Anaerobic digestion and the closely related collection of methane from landfill sites probably have the greatest potential world-wide. Other forms of biological waste treatment can provide useful and often profitable solutions for specific problems. For example animal feed (SCP) may be produced from food factory effluent. Ethanol and other low cost bulk products may be made by fermentation of a variety of wastes.

Anaerobic digestion

Anaerobic digestion has been known and used in many countries for thousands of years. In 1814 Davy started working at the fertilizer value of raw and digested cattle manure and collected biogas in a retort under vacuum. Until recently, however, commercial interest in anaerobic digestion was limited due to prevailing low energy costs, particularly for natural gas (methane), large capital investment requirements and problems related to process stability and efficiency. Anti-pollution legislation has stimulated research and development, notably in terms of reactor designs and treatment of strong – i.e. toxic – wastes. Rising prices of natural gas and other energy sources have provided an added stimulus. The results have been large increases in productivity from 1 m³ to 10 m³ biogas per m³ reactor volumes per day. These plants are now approaching commercial viability, though in general the return on investment would not be found attractive by most investors without other incentives.

Products and utility

If carbohydrate wastes are subjected to anaerobic digestion the resulting biogas, in theory, should be equimolar methane and carbon dioxide. In practice not all the carbon dioxide is released as gas because it is water soluble and can react with hydroxyl ions to form bicarbonates. Hydroxyl ions are produced during the deamination of protein as a result of the formation of ammonium hydroxide; thus protein in the feed will result in wastes with a higher pH and gas of higher methane content. Usually the gas varies between 60 and 80% methane. Hydrogen sulphide is also generated from sulphates and sulphur-containing amino acids in the feed.

In its crude form biogas can be used directly for cooking, heating, steam generation or lighting etc. on site. It may also be used to power stationary engines for generation of electricity, be compressed into cylinders for use as a vehicle fuel, or fed into piped gas distribution systems. If the gas is to be stored or piped, hydrogen sulphide must be removed to avoid corrosion damage. If gas is to be compressed, removal of carbon dioxide and water is usually necessary. When purified, biogas is equivalent to natural gas or synthetic gas. Methane has a very low liquefaction temperature (*c.* −70 °C). It is cooled and liquefied on a large scale, but on the small scales of anaerobic digestion this would be very uneconomical. This puts methane at a severe disadvantage compared to propane and butane which can be liquefied near room temperature.

If the substrate is assumed to be glucose, the weight yield of methane is approximately 27%, but the energy yield is over 90% when the energy contents (Table 8.1) are compared. In practice however the complex nature of raw materials in anaerobic digestors and low conversion efficiencies result in gross energy yields between 20 and 50%. The net energy balance is smaller, and sometimes may even be negative because of process energy required for transport and preparation of raw materials and separation, purification and compression of the product. Anaerobic digestion does have an advantage over ethanol in that the product, a gas, escapes from the liquid phase without an energy input.

Anaerobic digestion also results in the upgrading of the residual solid waste or slurry so that it may be applied to land as a source of fertilizer or even used as feed for livestock. In many farm applications the savings made in fertilizer or feed may exceed the value of the biogas. Sewage works often use anaerobic digestion as a method of sludge disposal after separation because sludge is otherwise costly to dump. It can then be used as a fertilizer. The methane can be used on site for the digestors but is rarely sold or used elsewhere.

Feedstocks

The potential raw materials for anaerobic digestion vary greatly, from agricultural wastes, processing wastes, municipal wastes and distillery slops to purpose grown crops in aquatic, marine or marginal land environments. Straw and similar stem wastes from other crops such as rape, cassava, sisal etc. constitute the largest source of plant waste matter arising in agriculture. They have a relatively low moisture content, a high fibrous content and can only be degraded slowly in anaerobic digestors. Other recyclable wastes such as potato and sugar beet have a higher moisture content and are more amenable to digestion. The effective digestion of both groups presents considerable technical problems; for example extensive chopping or milling may be required and lengthy holding times are necessary to achieve conversions, thereby increasing capital costs. Some dry fermentation systems have been developed but they are in the early stages as regards economic feasibility. The long retention times necessitate the use of large reactors, and low cost construction is essential for economic viability. There is also the question of alternative uses for these wastes, such as composting for animal feed.

The anaerobic digestion systems which are most common and most advanced are those which use manure and industrial plant effluent. They are located on farms, sewage works, fermentation plants, alcohol distilleries and dairies. They have higher moisture contents, lower retention times and therefore lower capital costs than the plant residue systems, but in some cases pre-treatment of the feed is necessary to reduce toxic effects. This occurs for example with high sulphate wastes, which would generate lethal concentrations of hydrogen sulphide in the digestor.

The potential contribution of anaerobic digestion to overall energy needs is very significant. The total quantity of biomass energy that could be generated from all animal and crop residues in the US has been estimated at between 1 and 10% of total US needs (Hayes *et al.*,

1980). The utilization of animal wastes, crop residues and unused forestry by-products is expected to fulfil 3% of the 1980 EEC energy demand by the year 2000 (Colleran *et al.*, 1982).

Digestion of human faeces, animal manure and crop residues has been traditionally carried out in small rural digestors in many countries. On a national scale the most significant contribution to a country's overall energy budget occurs in China, where there are over 7 million village-type biogas plants and several hundred larger biogas fuelled electricity generating stations. Several different designs exist but their common feature is that they are brick built with the gas holder and digestor combined in one unit built underground. The volume is about 10 m³ with no moving parts. Several thousand digestors have also been built in India, Africa and the Far East.

Farm digestors

The economics of anaerobic digestion are best illustrated by reference to cost schedules for individual operations. Some are estimates, but there are some available data on working systems.

In the first example projected costs for a digestor treating farm (piggery) wastes in the UK are shown in Table 10.1. They show an annual operating loss of £3298 for the instalment of this type of digestor. The greater part of the expenses (75%) is in repayment of capital. The running costs are surprisingly low, provided presumably that the system runs without problems. The volumetric capacity of this unit is not quoted, but by comparison to their other data it would appear to be approximately 300 m³. The authors point out that the figures do not include the value of residues produced after digestion and the reduction in slurry handling costs. They estimate the former at £5000 p.a. in replacing commercial fertilizer and the saving in slurry handling at £3650 p.a. To this is added an annual increase in electricity charges of 10% to transform the results into an annual operating profit of £8724.

Such calculations can be questioned on a number of grounds. Firstly in many countries electricity cannot be sold to the grid. Therefore unless all this energy can be used within the farm throughout the year the electricity benefits will be lower. Secondly the benefits of slurry handling costs and fertilizer replacement require careful examination, and with capital investment being such a high proportion of total costs one has to beware of overruns. The £50000 installation cost is very critical. Also electricity costs are at present not rising at 10% p.a. in most countries and it is dangerous to base investment decisions on projected rises due to inflation over 10 years.

Nevertheless the data do give an idea of the construction and operating

Table 10.1. *Annual cost for a digestor producing electricity and costing £50000 (5000 pigs)*

	(£)
Amortization	
25% of £50000 over 5 years	3638
75% of £50000 over 10 years	7200
Repairs and maintenance	
3% of capital (longer-lived) i.e. 3% of 75% of	
£50000	1125
Engine	1600
Lubrication	500
Generation	200
Boiler	100
Pump and meters	200
Instruments	200
Supervision	854
Total cost	15617
Power generation (100% utilization):	
350 days × 24 h × 35 kW at £0.0419 kWh⁻¹	12319
Annual loss	**3298**

From: D. A. Stafford & S. P. Etheridge (1982). Farm wastes, energy production and the economics of farm anaerobic digestors. In *Aerobic Digestion 1981*, ed. D. E. Hughes, D. A. Stafford, B. T. Wheatley, W. Beader, G. Lettinga, E. J. Nyns, W. Vestrate & R. L. Wentworth, pp. 255–67. Amsterdam: Elsevier.

costs of a farm digestor and how the economics depend on capital costs, digestor efficiency and the utility of methane. If cheaper digestors can be constructed or present size digestors can perhaps double their throughput or the utility of the gas can be increased, investment in this process could prove profitable.

In Israel the Kibbutz Industries Research Institute have designed and operated a farm scale anaerobic digestion system, the 'NEFAH' process (Marchaim *et al.*, 1982). They estimate capital cost of construction of a 200 m³ digestor system at $240000 (1981), assume a 10-year life span and put annual maintenance costs at 5% of equipment cost ($5300 p.a.). As with the previous example the economics of the operation depend to a great extent upon the utility of the digested slurry, which they claim can be used as an animal feedstuff. At zero credit for slurry the gas, which is 60% methane, 40% carbon dioxide costs 25¢ m⁻³, which is equivalent to 39¢ m⁻³ natural gas and is very expensive. In the best case analysis, with full utilization of slurry at $6 m⁻³, the gas can be produced for 5¢ m⁻³ or 7¢ m⁻³ natural gas equivalent, which is very attractive. As in the previous example the operating costs are low but capital costs are high – very high in this case. Why this plant should cost over twice as much as the UK unit is not clear from the available data, but it does seem that a more detailed analysis of construction costs has been made.

A third example of farm operation of an anaerobic digestor provides the most accurate and reliable information, based as it is on over a year's operating experience. This is the installation and operation of a 150000 US gallon (600 m^3) digestor on a farm with 1000 head of beef cattle in Illinois (Schellenbach, 1982). Here the installation and construction cost was $150000 (1980 prices), although much of the labour was provided by the farmer. The total operating costs for the first year were $10756, of which approximately $7000 was for electricity: 76% of this was required to operate the compressor to take the biogas from low pressure storage through a hydrogen sulphide purifier and store it at up to 250 lb in^2 in a 30000 gallon high pressure gas tank. Once stored in this way the gas can be used to replace propane, for example in the farm's ethanol-producing operation and in grain drying. Solids from the digestor could be recovered and used for animal feed, replacing silage at $35 ton, although additional labour and investment would be required in this operation.

In its first 15 months of operation the unit produced far more gas than the farm could use for heating and grain drying. Actual production was 21000 cubic ft per day which is sufficient to supply energy for up to 50 homes. If all the gas could have been consumed on the farm and replaced propane its value was put at $56000 p.a., but in fact savings of only $5000 p.a. could be made. The problem is therefore of finding uses for the gas. The intention was to run a farm ethanol plant using the biogas for distillation, but this had not become operational during the period of the study.

A number of alternative fuel uses and their economics for the digestion system just described are listed in Table 10.2. All figures are projected, including the propane replacement system actually operated.

The electricity generation examples require extra capital investment in provision of a generator and the vehicle fuel case needs an additional compressor to store gas at 5000 lb in^2 pressure and scrubbers to remove the carbon dioxide. The electricity figures assume that the local utility will purchase all the electricity generated at either 3 or 5¢ per kWh. Under these conditions all the alternatives, apart from the low cost electricity, are attractive, with good returns on investment or short pay-back periods. However there are a number of reservations. Firstly the farmer must become skilled at operating complex systems, and the figures all assume very low breakdown rates. The generation of these volumes of gas and usage on the farm are confined to certain types of

Table 10.2. *Comparative fuel use economics of 150000 gallon anaerobic digestion system installed on an Illinois farm*

	Installation cost ($)[a]	Return on investment (%)	Pay-back (years)
Natural gas	121000	18.8	4.8
Propane	150000	38.3	2.8
Electicity at 3¢ kWh⁻¹ (no credit for waste heat)	185000	4.3	8.9
Electricity at 5¢ kWh⁻¹ (no credit for waste heat)	185000	15.3	5.5
Vehicle fuel	210000	33.7	3.1
Proposed feed at 65% recovery[b] as silage at $25 ton⁻¹	156000	41.1	2.6
Proposed feed at 85% recovery[b] with chemicals at $25 ton⁻¹	158000	40.6	2.7

[a] All figures are projections.
[b] Assumes gas is used as propane but additional income is generated by recovery of the solids for feed.
From: Schellenbach (1982).

farming operations. Here they are collecting manure from 1000 head of cattle being fattened for beef in large barns. The farm also has 1700 acres of corn and soy beans plus 800 acres of grazing land. Operations on this scale and of this type are rare outside the United States. Only if large amounts of grain are dried can the propane gas return be achieved. Alternatively the digestor must be sited next to some energy-consuming operations such as an ethanol distillery. For smaller farming operations the capital costs will increase in proportion to the benefits obtained.

Industrial systems

The introduction of increasingly stringent environmental regulations has forced many industrial producers of high BOD or 'strong' wastes to look to anaerobic digestion as one method of waste treatment. An example is a 3.5 million US gallon (13–14000 m³) plant constructed by the Bacardi Corporation at their rum distillery near San Juan, Puerto Rico (Szendry, 1983). Rum is made by fermenting a solution of cane molasses with yeast and distilling the resultant beer. The residue from the distillery, termed slops, is a concentrated solution of impurities carried over from the molasses, containing high sulphate levels and many recalcitrant compounds produced from heating stages during recovery processes in the refining of sugar. The slops contribute over 95% of the total BOD of effluent from the distillery.

Several disposal techniques were considered, but anaerobic digestion was adopted after initial feasibility studies had indicated that over 70% of the BOD could be removed and the methane-rich gas utilized in the plant to replace a portion of the fuel oil purchased. Development work produced a fixed film matrix inside the reactor, giving a large surface area and downflow filter techniques. A full scale plant was constructed in 1981, incorporating several novel techniques, at a capital and installation cost of $8 341 000. Although it only treats half the effluent from the plant, much of the equipment has been sized and installed to accommodate the total effluent.

In operation the system removes up to 90% of the BOD and 75% of the COD of the waste. It produces 4.5 m³ gas per m³ reaction vessel per day, but the gas is only 50–55% methane because the effluent (a low protein waste) has a pH of 7.2–7.5. The gas can be used directly in boilers. In terms of gas produced the 1981 operating costs were between $1.50 and $1.75 per million BTU. Electric power costs followed by labour and chemicals were the largest operating costs. The economics of the plant are described as marginal at best. As the plant (at the time of the report) was not operating at maximum efficiency a pay-back time could not be calculated. The system could pay for itself but only when sewer use fees (i.e. environmental charges) are taken into consideration. The economics would of course be improved by rising fuel prices.

Landfill

The largest systems generating biogas are also the simplest in construction. They are the landfills used mainly for municipal garbage. The possibilities of using methane generated from such dumps has evolved from the need to overcome problems of fires and explosions resulting from gas evolution from such sites. Pioneering work by Los Angeles sanitation and public works departments prompted feasibility studies of conversion of landfill gas to electrical energy by the US Department of Energy. The continuous demand for electricity makes this the best and most cost-effective utilization, but it has only become feasible in the United States because of legislation that utilities must accept such power and pay a fair market value based upon avoided costs. It also means that economic returns on landfill gas are dependent upon the cost of electricity generation to the regional generating authority.

There are a limited number of landfill sites where the gas may be used locally without treatment. For example large quantities of domestic refuse are dumped in the pits resulting from clay extraction for brick manufacture in Bedfordshire, England. The clay soil is impermeable to the gas, the pits can be sealed after filling and the gas used directly to fire kilns in the nearby brickworks. It does not have to be piped or stored, and therefore cleaned up. Such opportunities however tend to be relatively rare.

Summary

Anaerobic digestion of agricultural, industrial and municipal wastes is of positive environmental value, since it can combine waste stabilization with net fuel production and allows the use of solid or liquid residues as fertilizer, soil conditioner or animal feed. The capital costs of anaerobic digestion are high, but the operating costs are low. In part the capital cost position may be improved by allowing longer write-off periods on the components, such as the reactor itself, which have life expectancies well in excess of the 10–15 years normally imposed by accounting practice. Capital charges also depend on the discounting rate or the expected rate of return, which in many cases may be set too high for this type of project. The capital costs may also be reduced by new methods of construction and new reactor designs. Improvements in reactor design are gradually reducing residence times and increasing gas productivity per unit volume. There is now a profusion of designs, including recycle systems, upflow filters, downflow fixed film reactors and fluidized bed reactors. Many resources are being devoted to increasing the surface area within the reactors without causing fouling problems. One problem with anaerobic digestion is that it is very dependent on the precise nature of the feedstock. Even slight changes in composition can cause operational blockages. At present systems have virtually to be individually designed for any one application, which confers high research, development and commissioning costs.

The value of biogas is marginal because of its content of hydrogen sulphide and carbon dioxide. A few locations such as the rum distillery and brickworks quoted can utilize the gas effectively and continually by direct burning, but these instances are in a minority. The costs of purification for storage and compression detract from its attractiveness in many uses. Generation of electricity is usually only feasible if it can be sold to the grid, because as with direct heating there are few locations with a year-round constant demand for electricity. Generating authorities are understandably not enthusiastic about accepting small quantities of power from individual suppliers. Many of their arguments concern the reliability of supply. If generators or switching gear should fail, the utility, not the private operator, is responsible for making good the supply to the grid. Legislation on acceptance of power as in the United States will probably be necessary elsewhere to gain growth in this usage. The operation of generators using biogas is also expensive because they require high replacement rates of lubricants.

In farm applications the economics of anaerobic digestion are dependent upon solid waste utility because energy requirements are highly seasonal (grain drying for example), and storage or treatment costs are not justified by the volume of gas produced, except in large intensive livestock-rearing units. In industrial applications there are now reports of successful systems, but cost analysis is difficult to obtain. Often the economics are dependent on waste disposal charges or penalties imposed by the local authority. Gas can often be used directly for heating or steam generation, but the capital and installation costs are still not covered. Anaerobic digestion is frequently only profitable at present if biogas is charged at zero cost, i.e. the facilities are already present or needed, as in sewage works or for environmental reasons. Under these circumstances it can pay to remove carbon dioxide and hydrogen sulphide. The future is dependent on fuel price trends and process improvements.

Given that the economics are often not favourable to private investors in Western societies there is the question of desirability of government investment in cost–benefit terms. Government might be prepared to expect lower rates of return if a net positive environmental and energy benefit can be demonstrated. In agriculture, for example, other forms of subsidy on equipment, product pricing and research are the rule, not the exception, in most countries. Providing schemes are well designed with a clear positive energy balance, there is a strong case for funding anaerobic digestion in such a way that operational units are established and that there are sufficient economic incentives for manufacturers to develop designs and innovate. Further improvements of the magnitude already achieved could reduce capital costs to an acceptable level and provide a useful source of energy to many communities.

Single cell protein from waste

Some wastes have a higher potential utility than can be realized by anaerobic digestion. One opportunity is fermentation to low value bulk products such as ethanol or single cell protein (SCP) which are extremely sensitive to raw material costs (Chapter 5). Although the energy balance for the ethanol fermentation is generally poorer than for anaerobic digestion, ethanol as a liquid fuel can command a much higher price on an energy basis than biogas. SCP processes in Western countries are not viable at all unless substrate is charged at a very low price. For the most part these processes have been operated on carbohydrate wastes. Although protein or hydrocarbon wastes are available they tend to present more technical problems or have higher value alternative uses. As with anaerobic digestion, environmental factors play an influential role in the overall viability of the process.

In one illustrative example, a confectionery manufacturer, George Bassett Holdings of Sheffield, UK, was faced with higher acceptance charges for high BOD/COD wastes from the local water authority. The factory produces *c.* 4 te per day of carbohydrate waste, the majority of which is concentrated in streams, totalling 140 000 l per day with COD

Table 10.3. *Production of SCP yeasts from confectionery effluent*

Analysis of confectionery waste solids (% w/w)		Annual operating costs for effluent treatment	Yeast process	Trickle process
			(£)	(£)
Sucrose	55	Power	26000	9000
Glucose	16	Nutrients, pH etc.	15000	14500
Starch	22	Bags	500	
Gelatine	3.5	Sludge disposal	—	27000
Caramel	2	Labour (£2 per man h)	4000	1000
Organic acids	1			
Coconut	0.5	Total cost	46000	51500

COD removal from effluent	COD mg l⁻¹	% COD removal	COD/BOD
Pre-fermentation	33827 ± 2548	74 ± 4	1.4
Post-fermentation	7795 ± 1353		1.7

Costs of COD removal (£)	Yeast process	Trickle process
COD produced (t p.a.) in manufacturing process	1612	1612
COD removed (t p.a.) by treatment	1198	1612
Process operating cost (£ p.a.)	46000	51500
Water Authority charge for treating residue (£ p.a.)	14500	
Total cost of treatment (£ p.a.)	60500	51500
Cost te⁻¹ COD (£)	37.5	32
Return te⁻¹ COD (£)	48	
Overall cost te⁻¹ COD profit (loss) (£)	10.5	(32)

From: A. J. Forage (1978). Recovery of yeast from confectionery effluent. *Process Biochem.*, **13**, **(1)**, 8, 11, 30.

values of 30–40000 mg l⁻¹. The remainder is found in very dilute streams, which cannot be treated effectively and which must be discharged with incurred costs. Over 70% of the waste comprises glucose and sucrose, and the remaining polysaccharide can also be readily degraded into fermentable sugars (Table 10.3).

Two schemes for disposal of this waste were considered; a conventional trickle filter effluent treatment process, and a continuous yeast fermentation plant using a food approved strain of *Candida utilis*. The costs are compared in Table 10.3. The yeast costings are based on pilot plant operation, the trickle filter costs on estimates. No capital cost estimates are given, other than that the yeast process requires 80% of the capital investment and one-tenth the land area of the filter system to treat the same effluent. The fermentation process comprises sterilizing the waste streams, continuously feeding a 200001 fermenter, collecting cells by centrifugation, drying and bagging the yeast. The most expensive items of the fermentation plant are a centrifuge to collect cells and a dryer. Together they account for approximately 50% of the capital investment.

The annual operating costs for the two systems are similar but have different breakdowns. Sludge disposal is the major cost in the filter system, whereas power for gas transfer into the fermenter and drying is the largest component in fermentation costs. The fermentation plant only removes 75% of the COD, the remainder being discharged into the sewerage system and incurring additional water authority disposal charges. The filter system removes all the waste and as a result has a total annual cost 20% below the fermentation plant. Overall however the fermentation system wins, because of sale of the SCP product. The plant can produce approximately 1.5 te per day or 516 te p.a. of a high grade product which has a 50% protein content and is rich in vitamins. It is suitable as animal feed, particularly for poultry, and was sold at £150 te^{-1} in 1978 grossing approximately £77000 p.a., a return of £48 per te of COD. The fermentation plant can therefore achieve earnings of £17000 p.a. as opposed to a cost of £51000 p.a. incurred with the trickle filter system.

As a result of these cost comparisons a full scale fermentation system was constructed and it has been operating successfully since 1979. It is both an example of a biological waste treatment scheme being the most economically attractive alternative through production of a saleable by-product and of the successful operation of an SCP plant when using a waste product. Unfortunately its significance is still rather limited. The factory produces large quantities of a high concentration waste stream comprised of almost entirely metabolizable carbohydrates and on a year-round basis. If any of these conditions were not met the SCP system would almost certainly be less viable than a filtration or sludge bed treatment. It must also be noted that without the water authority charges the plant would not be viable as it gives a negligible return on a capital investment of several hundred thousand pounds.

Potential and problems of whey

Whey is a good example for illustrating the difficulties encountered in the utilization of an apparently attractive waste material. It is the residue which drains from the curd (coagulated casein) in cheese manufacture and is composed of 93% water, 0.7% protein, 0.3% fat, 4–5% lactose and 0.5–0.6% salts, with slight variations depending upon source. Ten litres of milk are required for the production of 1 kg cheese with 9 litres of whey obtained as a by-product. World cheese consumption is approximately 7–8 million te annually, giving 4–5 million te of whey solids. Cheese consumption is rising, with particularly high growth rates in the United States (5% p.a.) and France (4% p.a.). Production is greatest in Western Europe (4 million te in 1982) and the US (2.2 million te). Approximately 50–60% of whey is utilized, at present primarily in animal feed and processed food such as ice cream, but the remainder, totalling almost 2 million te of lactose, is a waste, presenting a disposal problem to cheese manufacturers. It has been estimated that 1000 US gal per day of raw whey (the output of a small creamery), if discharged into a municipal sewage system, can impose a load equal to that generated by 1800 people.

This product could provide a useful resource as a feedstock for biotechnology industries, having the potential to supply 1 million te of ethanol for example or to satisfy the raw material needs of present citric acid or monosodium glutamate production. Unfortunately lactose is either non-metabolizable or only poorly metabolizable by most organisms used in commercial fermentations. It is also poorly soluble, crystallizing out of solutions above 20% concentration, and cannot therefore be transported and traded as 70–80% syrups with minimal microbial decomposition as is glucose or molasses. The output of whey is high during summer months, when adequate pasture produces a surplus of milk which can be converted to cheese, but is generally lower during winter. This tends to mean that the unsaleable surplus accumulates more in summer as the demand for the remainder is approximately the same year-round. Creameries also tend to be relatively small scale operations: a 5–10000 te p.a. production of whey solids is large, with the median in the USA being approximately 2500 te p.a. They are scattered through dairying regions to reduce milk transport costs, so collection and transport of whey, particularly in its undiluted form, present another cost. The anti-pollution legislation of the type encountered by the confectionery manufacturer has tended to be less of a problem to creameries because of their size and rural location. They are also often able to discharge it directly onto farmland to avoid polluting watercourses. In general creameries do not face the financial penalties at present to create pressure for effluent treatment, although this may be changing. Whey then presents almost a classic waste problem: it has very low utility, it is dilute, its production has seasonal fluctuation, and it has high collection costs.

Nevertheless there are a number of processes applied to whey, other than simply feeding it directly to livestock, which increase its value. The protein for example is a high value commodity in foods where it can compete with skimmed milk. Some whey (16% in the US and 3% in the EEC) is now subjected to ultrafiltration, which separates all proteins above 10000 daltons molecular weight, but the residue, whey permeate, still presents a disposal problem. Whey is also spray-dried to a powder, but this costs in the region of $100 to $110 per te drying costs alone (mostly energy), making it poorly competitive with many cereals and limiting its applications to some human food formulations and specialist animal feeds, such as calf-starter diets. Lactose extraction is another well-established use for whey, with the mother liquor from the extraction being dried and used in animal feed. Uses for lactose, which include pharmaceuticals, are however fairly limited and little market expansion is likely.

There is an extremely large volume of literature on potential biotechnological processes for whey utilization. Of many schemes proposed and often supported by optimistic costings, for example SCP production and lactic acid fermentation, only two – direct fermentation to ethanol, and hydrolysis of lactose to glucose plus galactose – have entered commercial operation, although in addition lactose is used as a substrate by some penicillin manufacturers. The fermentation of lactose in milk to ethanol by *Kluyvermyces* and *Candida* yeast strains has been practised for centuries in the production of beverages such as kefyr. Whey has only been used as a substrate for potable alcohol in recent years, for example the Carberry process in Ireland (Chapter 4).

Direct fermentation

Lactose fermentations are slower than the fermentations of glucose, maltose or sucrose, and the high yielding, fast-fermenting ethanol-tolerant *Saccharomyces* strains are unable to utilize lactose. They can utilize glucose plus galactose but they always show classical diauxie, utilizing glucose first and galactose only slowly, after having exhausted the glucose. The lactose-fermenting yeasts show lower ethanol tolerance than commercial brewery yeasts, so fermentations must generally be terminated at 1–2% lower ethanol concentrations. This results in higher reactor and distillation costs. Even so the predominance of substrate costs in ethanol fermentations will favour any apparently zero-cost raw material. Whey is not actually at zero cost, unless it is charged as a disposal problem; it must be concentrated to 15–20% lactose before addition to the fermenter, it must be heat-treated (if not sterilized), to reduce high bacterial counts, and it must be transported to the distillation plants. As ethanol fermentation economics favour large scale plants, particularly in the United States (Chapter 8), this is a considerable drawback. The combination of these factors has until recently precluded the use of whey in non-potable ethanol fermentations, particularly when compared to production from petroleum. The recent revival of fermentation ethanol has rekindled interest in whey as a feedstock, notably in the United States, where the federal tax exemptions (and in some states such as Kansas, Wisconsin, Iowa, Illinois and Minnesota, state tax exemption) apply to whey as well as to corn and can reduce the effective cost of ethanol by up to

$1.00 per gallon. A variety of strategies have been advocated to overcome the inherent economic disadvantages of whey, though it is not clear at present how many are in operation. For example, whey permeate can be concentrated first by ultrafiltration, then by evaporation to 40% solids and transported hot directly to the fermentation plant where it can be mixed directly with other medium components. It is almost impossible to cost this type of operation without knowing the distance between the plants, trucking costs and the basis of the agreement between the cheese manufacturer and the fermentation firm. The concentration costs will be a minimum of $20–30 per ton dry solids, which will be 80–90% lactose. The additional handling, transport and fermentation costs may undercut corn costs (even when corrected for by-product credits), but it seems unlikely to be by a large margin. It can at best be undertaken by a corn plant on an opportunistic basis. A custom built whey fermentation plant for fuel ethanol would need to be situated in close proximity to a number of large creameries to gain necessary economies of scale, and would have to be based on a long term supply agreement between creameries and the fermentation firm, which may involve consideration of waste disposal costs.

A single cell protein plant based on whey would have similar economics to the Bassett confectionery waste plant, since the feed streams would be of similar sugar concentrations. Again to benefit from economies of scale the plant would need to be supplied by several large creameries with the attendant transport costs. Unless there are very large rises in animal feed prices (in which case the value of the whey itself will also rise), these plants can only be economic if they offset large waste disposal charges.

One way of circumventing the whey problem has been proposed, at least for farm scale ethanol plants (Gibbons & Westby, 1983). This is to replace the water normally used in corn mash with whey. This permits a 20–30% reduction in the amount of corn required to produce a 10% ethanol beer using *Kluyveromyces fragilis*. Taking into consideration all credit and cost factors, the total annual cost of producing ethanol was $1.86 gal^{-1} using *S. cerevisiae* on corn alone and $1.59 gal^{-1} using *K. fragilis* on corn–whey mixtures (1981 prices). These costs were based on corn at $3.00 per bushel and whey at zero cost using small scale (under 1 million gal/year) farm plants. The whey savings seem optimistic when the fermentation times (114 h for whey *versus* 48 h for corn) are considered. This has been considered to give only a 10% increase in capital costs. A third option, *S. cerevisiae* plus hydrolysed whey–corn mixtures, still had a fermentation time of 90 h because of diauxie and was the most expensive option at $1.91 gal^{-1} because of enzyme costs.

Enzyme hydrolysis

Hydrolysis of lactose to glucose and galactose has the advantages of increasing sweetness, solubility and utilization of the product as a fermentation substrate. Higher concentration sweet syrups can be produced which may be handled as molasses and can compete with molasses in some applications and with glucose syrups in others. Potentially they can provide a useful route for whey utilization, but the end application and therefore added value is critical in determining what process expense can be tolerated.

Acid hydrolysis of whey is possible, but requires high temperatures and a low pH. Batch addition of mineral acid causes caramelization and colour formation in impure syrups. The hydrolysis is carried out under mild conditions, giving a very slow reaction time (as much as 48 h) and consequently high reactor costs. Strong cationic resins offer process improvements and can be operated at high temperatures, but the whey requires expensive pre-treatment to remove protein and fat components which will foul the resin.

Enzyme hydrolysis of lactose in milk products has been operated for some time. The β-galactosidases (lactases) used commercially fall into two classes: the yeast enzymes with activities at neutral pH and the acid fungal enzymes. The yeast enzymes are used in production of specialist milk products, either to increase sweetness, as in chocolate milk, or in diet foods for lactose intolerant individuals. There are also possibilities for pre-treatment of milk used in cheese and yoghurt manufacturing. They are not used in waste treatment, but in increasing value of foodstuffs, usually in batch additions, although SNAM Progetti in Italy used an immobilized system for producing lactose-free milk for some years. The fungal enzymes which are important for whey treatment are moderately priced enzymes, at present costing around $150 per kg depending upon supplier. One kg contains 20000 units (1 unit will hydrolyse 1 μmol lactose per minute at 30 °C). In a batch application enzyme costs are obviously dependent upon reaction time, but approximate to $15–25 per te dry whey solids, making them more expensive than the amylases or amyloglucosidases and putting pressure on the development of immobilized enzyme systems. Enzyme cost is not the only factor however; in common with other immobilized systems reactor costs are reduced. An additional problem with whey is the likelihood of microbial contamination, which often demands a reduced reaction time and therefore again increased enzyme costs. The methods of lactose hydrolysis in whey and approximate costs are summarized in Table 10.4. Immobilized enzyme processes are the cheapest alternative, but there are a number of severe technical problems which must be overcome.

Table 10.4. *Available alternatives for lactose hydrolysis in whey*

Method	Comment	Cost/te dry solids ($)
Batch acid	Degrades the product	40–60
Acid resin	Needs expensive pre-treatment	10+80–100 pre-treatment
Batch enzyme	Microbiological problems	50–100
Immobilized enzyme	Fouling problems	20–100

These costs refer to total hydrolysis cost, i.e. enzyme plus reactor and labour costs etc., but not to concentration, transport or other charges.

Immobilized enzyme processes

A great deal of research work has gone into whey hydrolysis by immobilized β-galactosidase and several commercial systems are now available, for example Corning, Sumitomo, and Tate & Lyle, although as yet few large scale processes are operating. It is important to examine the process considerations. First, the whey must be concentrated, because to operate the enzyme system at the starting 6% solids means large reactor volumes. Reverse osmosis can be used to concentrate to 10% (some manufacturers claim up to 20%) solids. If not, evaporation from 10–20% is usually carried out, so the process can be run at maximum lactose concentration. Hydrolysis is then followed by evaporation to 65–70% solids, the minimum at which microbiological growth is inhibited. Whey permeates may be hydrolysed without pre-treatment, but whole whey must be de-proteinized, generally by filtration, precipitation or flocculation, to prevent column fouling and unacceptably short enzyme half-life. The whey must also be demineralized, both for taste properties of the final syrup and to prevent salts precipitation when concentrated. Methods depend on the degree of dimineralization required. In some cases 50% removal is sufficient, in which case electrodialysis would be used. For higher degrees of mineral removal, ion exchange has to be used and costs are higher.

A detailed economic analysis of the Corning process has been made (Pitcher, Ford & Weetall, 1976) which, although now somewhat dated, is a useful treatment of the cost-sensitivity of immobilized lactase. The data refer to *Aspergillus niger* β-galactosidase bound covalently to controlled pore glass or titania, having an optimal activity of 300 IU/g and a half life of 62 days (35 °C) in a fixed bed reactor, the temperature of which is raised continuously from 35 to 50 °C over a period of 559 days to compensate for loss of enzyme activity. The whey was first demineralized and ultrafiltered. The sensitivity of processing costs to the

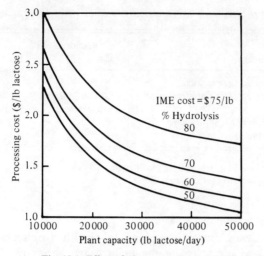

Fig. 10.1. Effect of plant capacity on whey hydrolysis processing
cost. From Pitcher, Ford & Weetall (1976).

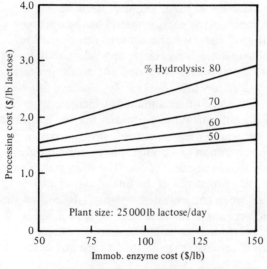

Fig. 10.2. Dependence of whey hydrolysis processing cost on catalyst
cost and conversion. From Pitcher, Ford & Weetall (1976).

degree of hydrolysis of lactose, immobilized enzyme cost and plant
capacity is shown in Figs. 10.1 and 10.2. Costs are not very sensitive to
the degree of hydrolysis up to 70%, but rise steeply above this level. If
the data were extended to include 90% hydrolysis they would show a
cost nearly double that at 70%. The average size of creameries
(20000 lb per day) is in a cost-sensitive region, so the economics of the
process are poor in small plants.

Table 10.5. *Treatment costs for whey permeate*

Process	$ te⁻¹ dry solids
Concentration from 6–10% (reverse osmosis)	20–30
Enzymic hydrolysis to 70% glucose + galactose	20–30
Concentration to 70% solids:	
Conventional three effect evaporator	70
Efficient evaporator with mechanical vapour recompression	40–45
Demineralization	
50% using electrodialysis	45
94% using ion exchange	110

Costs are only on an approximate basis for 5000 te dry solids per year plant and are sensitive to scale. No collection, transport, distribution or quality control/analytical charges are included.

The hydrolysis cost of the Corning process has been estimated to be in the $60–70 te⁻¹ dry solids range, although it may now be lower, with evaporation costs similar or perhaps less depending on the efficiency of driers (Chapter 6). When demineralization and transport costs are included the total process costs are probably in the region of $130–200 te⁻¹ dry solids. The process is aimed at producing a high grade syrup from whey permeate to compete with 60 DE (Dextrose Equivalent) glucose syrups in human food applications such as ice cream, confectionery or desserts. (DE is the total amount of reducing sugars, expressed as dextrose, that is present in a corn syrup, calculated as a percentage of the total dry substance.) It is a less attractive proposition in the United States than in Europe because of the lower glucose syrup prices, and as a result Corning have formed joint ventures with UK and French dairies.

The Corning process is still too expensive for products to be able to compete with molasses in animal feed applications or as fermentation substrates. Tate & Lyle have reduced hydrolysis costs to $20 te⁻¹ dry solids by using bone char, a cheaper immobilization matrix, and by being able to operate the process at 55 °C with enzyme half-lives in excess of 30 days. Even lower enzyme costs are claimed for the 'Lactohyd' process where fungal β-galactosidases are immobilized on Duolite ion exchange resins (Prenosil *et al.*, 1984). At the current low molasses prices however, the products of these processes will have difficulty in competing for most applications, unless they can attract a premium price. The major component in the cost is no longer the hydrolysis (Table 10.5). Evaporation, demineralization and possibly transport may all now exceed enzyme costs, a similar situation to that occurring with high fructose syrups (Chapter 7).

The economics of whey treatment now depend on reduction of de-watering costs, the development of higher value applications and the alternative disposal costs. If the claimed efficiencies of the new multiple stage evaporations with mechanical vapour recompression are substantiated in operation and the capital cost of this equipment can be justified, the process may look more attractive. Whole whey products fetch higher prices in animal feeds because of their protein content, so if protein removed before hydrolysis (say by flocculation) is added back, there may be a sufficiently high premium to justify operation of a process. Alternatively the future will rest, as elsewhere, with dumping legislation.

Cellulose and lignocellulose

The food industry produces numbers of dilute carbohydrate-containing wastes from processing plants which in general may be considered in the same terms as whey or the confectionery waste. Relatively cheap enzyme or whole cell systems can be used to overcome environmental problems and upgrade them, but the economics of the processes are contingent upon factors such as concentration, disposal charges, seasonality, capital investment and so on. While they present an opportunity for biotechnology, in global terms they are of very limited significance. Cellulose and lignocellulose wastes dwarf these wastes in volume terms in the same way that total lignin and lignocellulose dominate global biomass production, but they present much greater technical problems for treatment. The lignocellulose materials are paper and paper pulp wastes, wood (sawdust), agricultural wastes such as straw, municipal wastes and industrial wastes. Most of these materials contain cellulose, hemicellulose and lignin in the approximate ratio of 4:3:3. In western Europe and North America some 130 million te dry wt of wood p.a. are used for the production of pulp by chemical methods and a further 24 million te by mechanical methods. These processes generate some 65 million te of waste. On a world-wide scale this is approximately 90 million te annually. Theoretically, if convertible to a suitable feedstock, this waste could replace a significant proportion of oil used as a feedstock of the chemical industry. At present approximately 85% of wood pulp production is by the alkaline Kraft or sulphate process and 15% by the acid or sulphite method. Some by-products are obtained from Kraft pulping of highly resinous pine wood, such as terpenes and tall oil, but no industrial processes to upgrade the lignin and lignin degradation products are in operation. A large proportion of liquor is concentrated by evaporation and burnt to supply the energy requirements of the pulp mill. The spent liquors from the sulphite process containing sugars and organic acids are more

attractive raw materials. For example some 120 000 te of ethanol
p.a. world-wide are obtained from sulphite liquors by *Saccharomyces*
fermentation.

Feed-grade yeasts such as *Candida utilis* are also produced from
sulphite liquor. The residual liquors are concentrated and burnt. Both
sulphate and sulphite processes still have residual liquors which cannot
be concentrated or burnt and which must be treated in activated sludge
plants or aerated lagoons at significant investment and running costs.
Similarly the scale of straw wastes, such as corn stover and even
industrial or domestic lignocelluloses, is such that they could make
significant contributions to chemical feedstock and fuel supplies. The
annual total of agricultural wastes in the United States alone exceeds
800 million te dry solids.

Biodegradability

Cellulose is much more resistant to enzymic breakdown than
the sugar and starch wastes dealt with so far. It is also relatively
insoluble and often found in a crystalline form. The presence of lignin,
which binds cellulose and hemicellulose, makes the residue more
resistant to both enzymic and chemical attack. In economic terms this
may be expressed as high enzyme costs, both in terms of amount of
enzyme required and the slowness of reaction rates, which also imply
large reactor costs. The present rates are so slow that plants producing
the same tonnage of sugars as a present corn wet milling plant would
be orders of magnitude larger with capital costs that could not
conceivably be recovered. In order to facilitate enzymic degradation
these substrates are generally pre-treated by mechanical milling and
with acid, alkali or even radiation to break down polymers to
oligosaccharide levels. These processes too are expensive and have costs
which cannot be recovered in the sale of products at present market
prices. For example in a study on the production of ethanol from
agricultural residues (Sitton *et al.*, 1979), the authors concluded that
ethanol could not be produced for the then market price. In their cost
analysis acid pre-treatment of corn stover accounted for approximately
60% of capital investment, of which 40% was an electrodialysis unit for
acid recovery. In part the high costs were due to the need for
corrosion-resistant equipment. Operating costs were similarly
dominated by the pre-treatment steps, 40% again being associated with
acid recovery, even with corn stalks being charged at $15 per ton.

Other costs

Many wastes cannot be costed at zero because specific
collection costs must be incurred. For example straw or corn stover
collection is generally charged in the $15–20 per ton range. Current

estimates for domestic refuse in the United Kingdom put collection costs at around £20 per ton, although the bulk of this cost must be incurred anyway in sanitation terms. Even so additional costs are likely for the transport of domestic wastes to plants large enough for viability. Domestic refuse is also heterogeneous, even though a high proportion is cellulose. Some separation costs will therefore be incurred.

Production of agricultural wastes is also seasonal; if they can be stored, additional charges will have to be met. Although there has been progress in lignin and lignocellulose breakdown and the progress is accelerating, particularly in areas such as the enzyme systems of the white rot fungi and the fermentation of pentoses to ethanol, it is probably true to say that commercial realization is still some way off. Some wastes such as the sulphite waste liquors do present opportunities because the breakdown to metabolizable sugars has already been achieved as part of the main process.

Review of biological waste treatment technology

Biological methods can be used to detoxify wastes and in many cases can generate useful products. The most widespread technique, treatment of sewage on aerobic trickle beds, does not produce significant by-products and is just charged to local taxation. Anaerobic digestion upgrades the solid waste for use as a fertilizer or perhaps animal feed and produces a low grade biogas which must be used on-site continuously because purification costs are not justified at present energy price levels. It can be adapted to a very wide range of feedstocks and can make a significant contribution to global energy requirements on a local level. It is still a low added value treatment. Many wastes such as whey and certain food factory streams have potential as raw materials for ethanol or SCP and higher value products. If these processes can be commercialized the wastes will acquire by-product status. Lignocellulose wastes can make the greatest volume contribution to the industry and can supply a significant proportion of fuel and chemical feedstock requirements in many nations. Biodegradation of these compounds is still too slow for commercial application. They can only be utilized effectively when polymer breakdown has already been achieved, as in sulphite waste liquors.

All wastes have major disadvantages which limit their utilization. They are impure, often dilute, they tend to be seasonal, many cannot be stored and they frequently incur collection charges. In some cases the highest costs in biotechnological treatment processes are not in the actual enzyme or microbial breakdown, but in collection, purification or concentration. The processes must therefore be considered in their entirety and not just as biotechnological exercises. Unfortunately much research is often wasted on the biological step which is no longer rate-limiting. The most important restriction on the use of biotechnological processes is their capital cost. In most cases this is considered too high and the risks too great to justify the investment. In spite of much research effort and promotion many techniques can clearly be seen to be non-viable. The high capital costs arise from the requirement to handle large volumes of dilute waste, slow degradation rates imposing long residence times, or the general sophistication needed for a fermentation process. There is an urgent need for new reactor designs, methods of accelerating reactions and low cost construction to cut investment costs. Immobilized enzyme and cell systems with their low reactor and overall investment costs are already front runners but they still suffer from contamination and fouling in many applications.

The value of wastes is determined by their opportunity costs. Increasingly this is acquiring a high negative value because of dumping charges and penalties imposed by environmental legislation. This is proving to be the waste treatment industry's greatest asset. Plants are being constructed because of the need to avoid charges. Once operational they often prove successful and the knowledge gained can contribute to process improvements and reduced costs in subsequent systems.

Further reading

Biogas – fact or fantasy? D. A. J. Wase & C. F. Forster, (1984). *Biomass*, **4**, 127–42.

The prospects for biogas – a European point of view. E. S. Pankhurst, (1983), *Biomass*, **3**, 1–42.

Forster, C. F. (1985). *Biotechnology and Wastewater Treatment. Cambridge Studies in Biotechnology, 2.* Cambridge University Press.

11 New ventures in biotechnology – the impact of technology push

Historical background

The discoveries of monoclonal antibodies and *in vitro* recombination were widely recognized as being of enormous significance in biology and medicine and as having commercial potential. Applications of monoclonal antibodies as sources of pure antibodies to a single antigenic determinant in, for example, diagnosis were appreciated at an early stage. The development of cloning techniques gave rise to a wave of far-reaching speculation on their impact as well as concern over hazards, but commercial interest focussed on less controversial targets such as the bacterial production of insulin and growth hormone. Other eucaryotic proteins, notably the interferons, followed and generated intense scientific and commercial interest.

In economic terms the question arose of how this technology should be financed. The emphasis was of technology push rather than market pull or product diversification. Needs had to be identified and funding found. Furthermore it was realized that on an economic time scale the rewards would be slow in arriving. As emphasized in Chapter 3 economic planning is dominated by the discounted cash flow model. Relatively high research expenditure at an early stage followed by lengthy development, often involving regulatory approval, and finally the construction of expensive plant, are regarded with hostility by financial analysts and investors. Government and large corporations can afford this kind of expenditure and have contributed to biotechnology ventures. In some economies, notably that of the United States, development of this type is not seen as an area for government involvement. If new investment is to be attracted into this type of business, initial capital must be followed by injections of fresh money without income from products for some years. Confidence must not only be maintained, but increased – a momentum must be established. Early investors can then earn a return from capital gains. Finally, companies will probably have to make public share offerings to raise significant finance on acceptable terms.

Sources of finance

Internal funding

Until recently most commercial biotechnology has been funded by established firms as outgrowths of their business, usually in the same or similar markets, but using biological techniques. Some successful firms have grown from very small ventures, but relatively slowly. There has not been growth of the kind seen in the computer industry. Thus antibiotics were mostly developed by pharmaceutical manufacturers whose roots often lay back in medicines of the nineteenth and early twentieth centuries. Some food manufacturers diversified in a similar way and the large enzyme companies grew steadily, although relatively slowly. Chemical manufacturers have often used biological routes, where they were the cheapest alternative. In a rapidly evolving area such as genetic technology, established firms tend to have insufficient knowledge at first, or because of the nature of corporate decision making often move too slowly. New ventures find themselves unable to raise enough capital internally, unless they are in areas which are not capital intensive and/or have extremely high profit margins. Some suppliers of services and chemicals have managed to grow by internally funded developments and this has the advantage that the founders can retain control. However research expenditure and time scale of most biotechnological ventures have precluded the use of internal funding. Normal bank loans are also generally not obtainable because of lack of collateral in a new company and repayment conditions over too short a period. Many nations supply government grants, but these tend to favour projects in a more advanced stage, for example in plant construction where more employment will be generated.

Venture capital

Venture capital, also referred to as risk capital, is a type of direct investment in securities of new speculative firms or expanding technologically orientated firms. It is a form of long term financing, usually characterized as a high risk investment with large returns expected in dividends and capital gains. It has played an extremely significant role in funding new biotechnology companies and is worth looking at in some detail.

Investors in ventures cover the whole potential range: commercial or merchant banks, insurance companies, large manufacturing corporations, investment companies or private individuals. In the United States venture capital groups are classified as private or public. If public they must be licensed and come under federal regulations. Venture capital groups act as intermediaries between investors and entrepreneurs (Fig. 11.1), but they are not only suppliers of funds; they

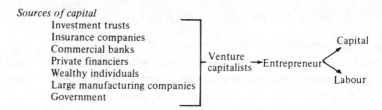

Fig. 11.1. Venture capital organization.

analyse the business and its prospects and actually function to assist the company. They will work with the company on its internal problems such as marketing and financial structure without taking direct control. Initially an entrepreneur or firm seeking capital is subject to rigorous evaluation. A detailed application, the business plan, must be submitted, describing the company's aims, its past record, projects, current balance sheet, personnel, patents, market analysis and so on. The venture capital group scrutinize this, paying particular attention to analysis of projected earnings, product selling price, market size, ease of entry into the market, technological advantages, production feasibility, quality and quantity of the competition and so on. The criteria are broadly similar to those outlined for internal project evaluation (Table 3.2). It has been estimated that up to 95% of all proposals are rejected in the initial stages of consideration and 3–4% after more detailed scrutiny. Only 1–2% ever receive final approval. Although venture capitalists will invest in projects ineligible for financing from traditional sources, they are often more cautious in screening than other financial institutions, for example commercial banks, in most loan situations.

The entrepreneur must also assess the investors if he has a choice. There are dangers of the investor's goal being simply to acquire the company and its technology. Venture investments are a type of partnership, but differ from the standard business partnership where each partial owner has a vote equal to his holding of equity. Firms financed by venture capitalists generally relinquish a greater portion of control to the venture capitalist, even though they may have the same equity holdings.

If the application is approved, venture capital may be supplied on an annual rate of return basis or on a portion of equity ownership. Most biotechnology ventures have been funded on the latter basis, which is more optimistic, with higher potential rates of return, but obviously carries more risk. Historically expectation of rates of return have varied with the amount of risk associated with the investment. In general, venture capital groups aim for minimum investments of $100 000 upwards because small projects take as long to evaluate as large ones. The firm to be funded must also be of sufficient size to make a sale, merger or public issue worthwhile because the venture capital firm's investment policy is aimed at achieving most of their profits through capital gains in the stock. The main problem for venture capital investors is that the investment is in a non-liquid form for an extended period of time. Numerous unpredictable events can adversely affect the investment and the investor cannot withdraw without losing heavily or completely. For example high interest rates or rates of inflation may make the investment unattractive in relation to alternatives. The significance of venture capital to biotechnological development can be seen in that in 1980 and 1981 biotechnological investments in the United States were attracting nearly $100 million of venture capital per year.

Joint ventures

In some cases venture capitalists will group as a consortium on a joint participation basis to finance large projects and spread risk. The British biotechnology venture Celltech is an example where several companies and a government agency have joined. In these arrangements companies may have equivalent roles or one investor may serve as the principal underwriter. Some large companies with sizeable research expenditures maintain an entrepreneurial-oriented subsidiary and many large companies establish a joint venture to give them access to the biotechnology or product of the smaller company. They supply capital and possibly marketing and production expertise. This is of value to many biotechnology ventures where the link in their product range is a common means of production, while the products themselves may be in widely differing markets. As a result many companies in genetic engineering have acquired corporate investment which is tied into the marketing rights of the products. For example Genentech and Eli Lilly have done so with human insulin, Schering Plough and Biogen with interferons. This has often been to the mutual benefit of both the smaller company, which has acquired finance on favourable terms, and the larger company, which has had access to the technology for less than perhaps its inhouse development might have cost. In general the corporate investor in these cases only has a minority shareholding.

Limited partnerships

Limited partnerships are a form of investment where investors buy an interest in a specific research project rather than in the company itself. Tax benefits can be gained because each partner can deduct all expenses for research from income in the year the expenses were incurred, provided the partners were 'at risk', and the company gains because it acquires interest-free capital without issuing new equity. The money is classified as contract revenue and not as a liability on a balance sheet. This method of financing has proved extremely popular in biotechnology, the industry having raised an estimated $350 million by it in 1980–83 (Lee *et al.*, 1983). It is a particularly important method for companies in health care fields to underwrite the very large expense of clinical trials necessary for regulatory approval.

Wealthy individuals

In the United States in particular, wealthy individuals paying a high marginal rate of tax are prepared to make high risk investments, particularly if there are tax advantages in doing so. Doctors have been a source of funds to small medical and biological companies. There are disadvantages to this method however, often arising from personal friction between the entrepreneur and investor.

Mutual funds

These are generally organized by investment banks to make high technology investment opportunities available to the small investor. They differ from venture capital in that the investment group merely buy into the industry, but do not supply post-investment management advice and support.

Public issue of shares

In general, capital may only be raised in this way after a company has already been built up to a reasonable level by one or a combination of the methods described. The first biotechnology companies went public in 1980, generally after a venture development phase of between three and ten years. Funds can often be obtained on more favourable terms than by other methods, but firms are open to public scrutiny and the vagaries of the stock market.

Scale of investment achieved

These various methods of raising capital have funded the development of over 200 biotechnology companies in the United States since the late 1970s. Cumulative investment was put at $1.87 billion in December 1983, of which $1.4 billion was in equity (public offerings and private venture capital) and $500 million in research and development limited partnerships (Murray, 1984). In addition to this there is expenditure by large corporations and governments. For example *Business Week* estimates a $2.5 billion total investment by the end of 1983, but this probably includes corporate spending. Of the total of $1.87 billion, $619 million was raised in 1983, $378 million in 1982 and $870 million up to 1982 (Fig. 11.2). In terms of market areas it was estimated that 72% was invested in health-care, 16% in agriculture and 11% in others (Murray, 1984).

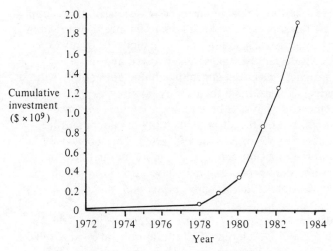

Fig. 11.2. Capital investment in biotechnology companies. Investment figures from Murray (1983, 1984).

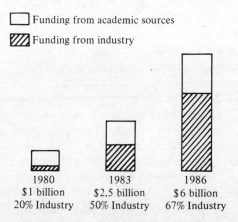

Fig. 11.3. Spending on biotechnology research world-wide. From Nelson Schneider, E. F. Hutton & Co. Inc.

These methods of funding are resulting in a dramatic shift of the funding of biotechnology from public to private sectors. It has been estimated that of total biotechnology research spending world-wide of $1 billion in 1980, 20% was by industry and 80% by universities and government research institutes. By 1983 the total was $2.5 billion, funded equally by industry and research, and by 1986 predictions are for $6 billion with 67% coming from industry (Fig. 11.3).

The success of the capital generation has enabled companies to fund many years of expensive research with very little being earned in revenue. In some cases fund raising has been so successful that, coupled with high interest rates, firms have been able to finance their research from interest received from their capital. The confidence of investors in biotechnology, considering the time spans involved and the slow appearance of products, has been enormous, but is not entirely without historical precedent. For example the railway boom of the mid-1830s developed a momentum that was based not on tangible evidence of commercial success, but on the sheer availability of capital and the hunches of businessmen (Dyos & Aldcroft, 1969). The successes of venture capital in the other burgeoning, but more mature, area of microelectronics and data processing may have also played a role in establishing confidence and encouraging speculation in another technology.

As an example, in 1957 a public venture capital company – the American Research and Development Corporation – invested $70 000 in a project called Digital Equipment. By 1971 the net worth of the Digital Equipment Corporation was $575 million (Dominguez, 1974). Experience in general in the United States has shown that venture pools in new technology have outperformed both stocks and real estate. In addition the US government reduced capital gains tax from 50% to 20% in 1978 to make capital growth attractive relative to dividends and it gives encouragement to high technology industry by allowing tax credits for increases in research and development expenditures.

There is now some evidence that investment patterns in biotechnology are changing. The contribution from venture capital fell in 1982 to *c*. $50 million and the number of new companies receiving support also dropped. There has been a rise in the popularity of limited partnerships but, more significantly, companies have been going public. In 1980 the first biotechnology company to go public, Genentech, offered its stock at $35 per share; these jumped within minutes to $89. This resulted in the value of the company leaping from $260 million to $660 million which was insupportable and later fell. (The company's original venture backing in 1976 was $100 000.) Since then approximately 60 biotechnology companies have gone public and their share prices have fluctuated because they are high technology, high risk stock. In particular there was a crisis of confidence in late 1982, although stocks later recovered. In general the US stock market has been bullish in the early 1980s, the market has been strong and there has been a willingness to speculate. Even so biotechnology stocks have outperformed the market average (Anon., 1982). If the stocks were to fall in price the companies would be prey for cheap takeover bids and large companies could buy the technology cheaply. The failure rate of biotechnology companies has been far lower than expected. There have

Table 11.1. *Summary of funds raised and expended by US biotechnology firms in 1982 and 1983*

	1982		1983	
	Amount ($ millions)	No. of firms	Amount ($ millions)	No. of firms
Initial public offerings	62.1	5	485–499	35
Secondary offerings or warranty	39.00	3	123.2	9
Joint ventures/partnerships	19.7	2	n.a.	n.a.
R & D limited partnerships	198.6	9	195.4	5
Private placements or loans	190.2	35	78.5	17
Total raised or committed	509.6	—	872.8–886	—
Total expenses	336.5	49	453 (estimated)	50

n.a. not available.
From: Anon. (1983). The biotechnology industry: real or imagined? *Biotechnology News*, 3, (22), 8.

been predictions of several shakeouts and an overall survival rate of 10–20%, but so far this has not occurred.

In 1983 the top 50 or so biotechnology companies spent an estimated $450 million (Table 11.1) and employed nearly 6000 people. Extrapolating from this it was estimated that the industry as a whole may have spent over $2 billion and may employ *c*. 30000 people. Even if these figures are exaggerated it is certain that the biotechnology industry is sizeable in terms of both spending and employment and that its expansion has been meteoric. In addition many large companies, notably in oil and chemicals, such as Exxon, Du Pont, Hoechst and Monsanto have formed contracts with US universities to fund research in biotechnology in return for exclusive rights to results for commercial exploitation. These contracts often involve large sums of money. The Hoechst-MIT collaboration is put at over $70 million over 10 years (Fishlock, 1982).

National strategies outside the United States

The explosive growth of American biotechnology has not been paralleled elsewhere. In part this is probably a result of the lead established in recombinant DNA technology by American universities, but in addition it is due to the confidence of American investors in high technology ventures, the systems for raising capital, the tax incentives for what are considered high risk ventures and the strengths of entrepreneurial motivations amongst Americans. The extreme wealth of many individuals also means that they have spare money, which would otherwise attract high marginal tax rates, to invest in risky ventures. Nevertheless other nations have instituted programmes in biotechnology, almost all government aided or government co-ordinated. In most cases they have recognized the role of private sector enterprise to develop new business areas, but have attempted to supply or supplement the capital which cannot be generated as in the American system.

Biotechnology has its strongest research base in the United States, but it has its strongest commercial base in Japan. Japan is a world leader in the production by fermentation of nucleotides, antibiotics, amino acids and enzymes. Gross sales from the exploitation of microorganisms probably reach $10 billion annually, but venture capital is virtually non-existent in Japan. Japanese industrial policy is co-ordinated by the Ministry of Trade and Industry (MITI). It has tended to support the development of strong, fiercely competitive oligopolistic companies in specified business areas. These organizations are most familiar in automobile manufacture, cameras, consumer electronics, motorcycles etc. A similar policy is being adopted in biotechnology, which has been identified as a target area for future growth. MITI has allocated $128 million in a 10-year programme of supporting research in selected companies in areas such as food, chemicals and pharmaceuticals. The research areas have been split into bioreactor development, mass cell culture and recombinant DNA (Fishlock, 1982). The recipients include many of the successful companies familiar to Western biotechnologists, for example Ajinomoto, Takeda, Asahi Chemical, Sumitomo Chemicals, Kyowa Hakko Kogyo, Mitsui, Mitsubishi Chemicals and Kao Soap. The companies also have their own internally funded research programmes. It is estimated that 150 companies are spending $217 million p.a. on biotechnology research and many have formed liaisons with Western biotechnology companies and operate joint research programmes. Japanese universities also have strong industrial links.

The United Kingdom has two larger biotechnology venture companies funded by a mixture of government funds via the British Technology Group and private investment from merchant banks, large companies, insurance and pension funds. Celltech, capitalized at about £30 million, concentrates on medical and diagnostic areas, particularly in monoclonal antibodies, although it has wide ranging minor research interests. The Agricultural Genetics Group, capitalized at £23 million, has eponymous interests.

In France also, venture capital is thin on the ground. In 1982 the French government earmarked $100 million for a biotechnology mobilization programme with the aim of allocating public funds to firms doing development work. The aim is to create at least five companies on the same basis as Celltech and Agricultural Genetics. Nationalized industries, for example Rhone Poulenc and Elf Aquitaine, are also receiving government aid to construct biotechnology laboratories. West Germany too has allocated some $90 million to industrial concerns and $40 million to universities to promote biotechnological development, and this spending is expected to increase. The European Economic Community has also allocated limited funds to date, but is expected to increase spending in this area.

Earnings

Scrutiny of the balance sheets of the publicly quoted biotechnology companies indicates that for most, expenditure has far exceeded non-equity income. Also most income has arisen from interest plus a limited number of licensing deals or selling of technology. Most companies have in fact financed their research and development through public stock offerings with only a few, if any, products on the market. Many companies face critical periods of their development when cash flow considerations, e.g. earnings from product sales, will be crucial. There was increasing evidence through 1984 that earnings from products were beginning to appear.

First generation products

Critics have remarked that there are more biotechnology companies than products. The validity of this statement depends on how finely a product is defined. There are now many known interferons, hundreds of monoclonal antibody lines and quite a large number of potential vaccines. Nevertheless there has been a concentration on a narrow range of market areas, nobably high value human health-care products and veterinary medicine. To some extent oversubscription is a predictable consequence of technology push, but there is a strong element of market influence. Health-care and veterinary products are the type of high margin, low volume products that can make a realistic return on expensive research expenditure without further capital investment in large scale plant. Veterinary products, although having perhaps slightly lower margins, win out in terms of their lower cost of toxicological approval.

The first generation products, as outlined in Table 11.2, reflect these considerations. Human insulin has been marketed since 1982 and is now gaining appreciable market penetration. Human growth hormone is now undergoing clinical trials, and its future seems reasonably assured.

The majority of products now emerging from biotechnology companies are diagnostics of human diseases or veterinary medicine based on

Table 11.2. *First products of biotechnology companies made by recombinant DNA technology*

Products marketed	Human insulin
	Several animal vaccines
	Monoclonal diagnostic kits
Advanced development/	Human growth hormone
marketing trials	Animal growth hormones
	Human urokinase
	Veterinary therapeutics
	More monoclonal diagnostics
	Human interferons, lymphokines, interleukins
	Hepatitis B viral antigens
	Animal viral antigens e.g. foot and mouth
	Calf rennin
	Tryptophan (animal feed additive)
	Phenylalanine (Aspartame manufacture)

either monoclonal antibodies or DNA probes. Monoclonal diagnostic kits for breast and prostate cancer, hepatitis B virus and Epstein–Barr virus are now available. These are all small applications; a mere 100 l capacity can supply a huge number of kits. Monoclonal antibodies for the purification (affinity columns) and assay of interferons are also being marketed. This is an area of rapidly evolving technology, with some of the DNA probes now appearing apparently better than monoclonal antibodies for some diagnoses, for example for herpes and chlamydia infections.

Other products in advanced stages of development include animal growth hormones and vaccines against scours and foot and mouth disease. On the fermentation side amino acid manufacturing companies are working on L-tryptophan and L-phenylalanine.

The development of all these products represents considerable technical achievements, notably in the speed of their development, but a realistic assessment is that they will not, in the words of one analyst, be blockbusters – the markets will be small. In general they do not justify the optimism and the scale of investment to date. To understand these phenomena it is necessary to look at market predictions for the medium and long term future of this technology.

Market forecasts

Predictions of markets for biotechnological products have two outstanding characteristics: they are expressed in tens of billions of dollars and there is variation according to source. For example total sales in biotechnology have been put at $550 million in 1990 and $50 billion in 2000 (Information Resources Ltd), $27 billion in 1990 and

$64 billion in 2000 (TA Sheets & Company), $40 billion world-wide in 2000 with a US total of $14.6 billion (US Government Office of Technology Assessment) or over $3 billion in 1990 from recombinant DNA products alone (International Resource Development). Diagnostic kits are predicted to be a $500 million market by 1990 (US Government Office of Technology Assessment), while monoclonal antibodies alone are projected at $1 billion by 1990 (Vaughan, 1982), with most sales in diagnostics. One difficulty is that most figures are quoted without detailed reference to what is being measured. Do total biotechnology sales include bulk products like ethanol and citric acid, or do they refer to the products of recombinant DNA plus monoclonal antibodies alone? If recombinant DNA techniques were to make even minor refinements to a process such as monosodium glutamate production, $500 million per year could be tagged onto markets.

Given the reservations about figures there is a great deal of optimism on the future for biotechnological products. They are widely assumed to become billion dollar markets over the next few decades, despite modest performance so far. There are a number of factors in this reasoning. First the major initial applications appear to be in health-care. As the gross national product of a country increases there is an increase in the proportion of GNP allocated to health-care. Furthermore these developments are at the high technology end of medicine where in the past firms have made good returns, particularly in affluent economies such as the United States. There is also an element of straightforward gambling. Interferons or lymphokines, if they succeed as anti-viral or anti-cancer agents, will have an impact equal to or greater than antibiotics. If they do not the market value for research use will be quite small. In the longer term the changes in the relative prices of agricultural commodities and petroleum promise big market opportunities of biological replacement of petrochemistry, and still greater improvements in agricultural productivity are possible. Recombinant DNA is viewed as the enabling technology. Again figures for market projects can be vast, but contain a high risk element. Genex for example estimate the market value of recombinant DNA products to be $40 billion in the year 2000, of which $25 billion will be in petrochemicals. To develop this business they estimate will require $24 billion, 80% of which will be required for capital expenditure for petrochemical facilities. To put this in perspective it is worth remembering that the present market of biotechnological products is approximately $20 billion (40% antibiotics, 20% ethanol, and 10% high fructose syrups). Present world sales of crude oil at $30 per barrel are approximately $600 billion.

The short term sales of the speciality biotechnology companies are forecast to grow in the $50 to $200 million range through the mid-1980s. This is against the total valuation of $3 billion of the top 20 publicly quoted companies in mid-1984 (Anon., 1984*a*). This is another example of the exceptional investor confidence in biotechnology ventures, for as a general rule of thumb the market valuation of fast growing companies is only 1.5–2 times their next year's sales.

Product categories

Expenditure in biotechnology companies has been dominated by human health-care. For example when equity investments in companies were broken down into their market areas, human health-care totalled 61%, followed by agriculture at 23% and others (mainly chemical processes) at 16%. A similar pattern is seen in research and development limited partnerships. The majority of the funds in agriculture were invested in crop improvement projects rather than animals (Murray, 1983). Health-care products tend to be favoured because there is a clearly defined path from laboratory experimentation to the market. A product, say a protein, can be purified, tested as a therapeutic, patented and marketed by tried and tested procedures. Its efficacy and market can be evaluated accurately and the returns earned can be very high. On the debit side the gaining of regulatory approval is long, costly and risky. Plant genetics is a much more poorly defined area. Less is known at the molecular level, there are few vector systems and perhaps less certainty that patents, if granted, can be policed. Nevertheless this is potentially one of the areas with the greatest economic impact.

The supply industry

The business of selling products to the biotechnology industry has been a strong growth area and is booming. Products include enzymes, synthetic oligonucleotides, filters, fermentation and downstream processing equipment. The present market of approximately $100 million is forecast to rise to $300 million by 1990.

Diagnostics

In vitro diagnostic kits have been a strong growth area since the early 1970s. To date they have mostly been based on enzyme assays and radioimmunoassays. Each diagnostic kit is sold as a package plus instructions, and the cost of the actual components (the enzyme or the radio isotopes) is usually a small fraction (< 10%) of the selling price. The total value of the *in vitro* diagnostic kit market is forecast to reach the $1–1.5 billion range by the early 1990s. The freedom from regulatory constraints has meant rapid commercialization for diagnostics.

The ability of antibodies to bind stoichiometrically to specific antigens at very low concentrations makes them valuable in diagnosis for detecting toxins, hormones, cancer markers, or bacterial or viral antigens. Monoclonal antibodies have a number of advantages over radioimmunoassays: they are specific for one antigen, there are no batch to batch variations or false positives from cross-reacting material. They should be cheaper to produce because cell culture techniques can be used in place of live animals. Monoclonal antibodies are being produced on the 1000 litre scale in airlift fermenters (Celltech), encapsulation in 100 litre fermenters (Damon Biotech) and in perfusion chambers using lymph from live cattle (Bioresponse). This plant, costing $3.5 million, can produce 30 kg per year of antibody at present prices of $1000 per g. The encapsulation technique requires 1000 litres of fluid to produce 1 kg monoclonal antibody, whereas conventional tissue culture would require 100000 litres. Also a portion of the cost of using radioimmunoassays is in disposal of the radioisotopes which is clearly eliminated with monoclonals.

There is a caveat in this development. The ease of entry into the business and the relatively rapid approval of *in vitro* diagnostics have attracted many newcomers, and the technical skills for developing hybridomas and monoclonals have become widespread; coupled with the lowered production costs, this has led to intense competition. By the spring of 1984 in the United States alone at least 100 companies were developing products based on monoclonals. Over 600 different monoclonal antibodies were available and about 60 *in vitro* diagnostic kits incorporating monoclonals had been approved (Anon., 1984*b*). Five companies were manufacturing monoclonal antibody kits for pregnancy detection. The situation is similar to that seen with radioimmunoassays in the mid-1970s when dozens of small firms were started. There were many failures, mergers and take-overs. In many cases sales grew rapidly to a few million dollars but then levelled off. As with monoclonals the very diversity means demand for each type tends to remain small. It is also a rapidly evolving field with other forms of competition; for example fluorescent antibody techniques are becoming popular (although monoclonals can also be used in association with fluorescent markers), and DNA probes are proving better in some detection applications than monoclonals. The diagnostic kit field is an interesting example of pure and perfect competition whereas other areas of biotechnology have tended to become oligopolistic.

Monoclonal antibodies have many exciting possibilities besides *in vitro* diagnosis in the medium and longer term. They may be used for *in vivo* diagnosis, for example in tumour imaging, and they may have therapeutic applications. For example they may be directed against allergic responses in autoimmune disease or transplant rejection, they may be used in tumour suppression, against antibiotic-resistant bacteria or parasites, or to carry drugs to a target antigen, such as a tumour cell. Development in these fields will be slower and more expensive with regulatory hurdles to be overcome, but already the possibilities of large scale production has enabled the use of monoclonal affinity ligands to purify expensive products such as interferons, interleukins and factor VIII.

Peptide hormones

Human insulin was one of the first targets of recombinant DNA research in the earliest development phase of the technology. There were two reasons for this: first, the number of human diabetics requiring insulin therapy was expected to outgrow the available animal (pig pancreas) supply, and second, human insulin might not produce the adverse reactions produced by the animal products in some patients. Market considerations caused companies to switch from making the new human insulin for two reasons. Firstly, Novo were synthesizing a product with the human amino acid sequence from porcine insulin by chemical modification of one amino acid, and secondly, physicians would be unwilling to switch existing patients to a new product if they were satisfied with their existing regime. In the event only one company, Eli Lilly, already a world leader in insulin manufacture, has marketed the human product made by bacterial fermentation (*E. coli*) using a process developed by Genentech, under the trade name Humulin. The product gained regulatory approval in 1982 and has been marketed in the United Kingdom since then. Of 300000 insulin dependent diabetics in the UK with a total market of £18–22 million p.a. 7000 (2%) were being maintained on human insulin in May 1984. Human insulin now has 15% of the Danish market, but it is not clear what the breakdown between the Novo and Eli Lilly sales are. Lilly have now priced their human insulin marginally above the porcine product in the United States. The total consumer value of the US insulin market is around $300 million annually (Tokay, 1983). It has been estimated that Lilly sold $25–30 million worth of Humulin in 1984 and it is predicted that sales will rise rapidly. It should be noted that both companies are market leaders matching a new product against their existing ones. The situation might be very different for a new company attempting to enter a market. Lilly has assured its supply should an animal pancreas shortage emerge and has secured a good defensive position against its major rival, Novo.

Growth hormone, also being a short polypeptide of known sequence, was one of the first target products of recombinant DNA technology. The need for it arises because one child in 5000 suffers from hypopituitary dwarfism as a result of growth hormone deficiency. This condition can be treated by administration of growth hormone in childhood. The only source of this hormone at present is extraction from the pituitaries of cadavers. Approximately 10000–15000 children in the United States are being treated by this method, which requires 30 pituitaries per child and costs an estimated $5000–10000 per treatment. The availability of a recombinant DNA product will allay the tenuous supply situation, reduce the costs of treatment and permit treatment of children who, while not seriously deficient, will benefit from growth hormone administration. It may also assist in wound healing.

Genentech have led the way in producing growth hormone and were hoping that their product would be given regulatory approval in 1985. The product development is assisted by easy marketing because only a limited number of physicians are targets, a substantial advantage for a company with limited resources and an expensive research budget. The annual market is estimated at $100 million world-wide.

Interferons and lymphokines

The promise of interferons as effective treatments for viral diseases and cancer has been the single most significant stimulus for investment in biotechnology ventures. Indeed there have even been attempts to correlate the fluctuations of stock prices of genetic engineering companies with publicity on interferon, although they have had only mixed results (Powledge, 1984).

Study and evaluation of interferons had been hampered since their discovery, because the tiny quantities synthesized by mammalian cells prevented scale-up to supply even laboratory demands, let alone realistic clinical applications. Recombinant DNA technology has overcome this hurdle with spectacular success and gram quantitites are now synthesized routinely. Unfortunately the effects of the first pure interferons to be made in large quantities have been disappointing. As anti-viral agents they only seem to have realistic applicability against severe, even life-threatening, infections since they themselves appear to produce many of the symptoms of less severe infections such as nausea, fever and headaches. They also seem to have limited efficacy as anti-tumour agents, particularly against solid tumours, although more promising results have been achieved against types of leukaemia. A single species of interferon is generally less effective than a heterogeneous preparation from human cell tissue culture, but tissue culture is much more expensive, with yields in the range of 10^8 units ml^{-1} *versus* 10^{10} units ml^{-1} from recombinant DNA bacteria. These are early days to pass judgement on interferon as a therapeutic agent. It may take years to develop the right mixture or combination with other drugs. Modified interferons, say by glycosylation, may have the same efficacy but fewer side effects. The side effects may be dose-related and so on. If not a panacea, interferons are still valuable products and the achievements of recombinant DNA technology in this field should not be underestimated. A dose of interferon either completely unattainable or with a price in the stratosphere 10 years ago can now cost as little as $30.

Fibroblast (α) interferon now has regulatory approval in West Germany for the treatment of genital herpes and life-threatening viral diseases, against which it has some efficacy. Its cost is put at DM 1000 (*c*. $400) for 5 million units.

Lymphokines, the protein immune products of lymphocytes (of which interferons are one class), are now thought by many to offer the most exciting possibilities of all the potential recombinant DNA products, but it is still too early to say how they will perform in clinical trials. They offer the possibility of enhancing or restoring the immune system to fight infectious diseases or cancer. Interleukin-2 is attracting the greatest amount of interest at present. The total research effect on lymphokines is now very large. As with interferons, recombinant DNA technology enables the synthesis of gram quantities of the product which is only made at the level of a few molecules per mammalian cell.

In commercial terms, perhaps, too many companies are dependent upon interferon. Both interferons and lymphokines can only be described in terms of being high risk gambles with multi-billion dollar markets if successful (Anon., 1984c).

Vaccines

Conventional vaccines usually consist of pathogenic bacteria or viruses which have been rendered non-virulent by death or attenuation. Genetic engineering permits the construction of specific antigens conferring immunity without the risks associated with culturing pathogens or of reversion of attenuated strains. Products could be purer and have more reproducible effects. They may also permit synthesis of antigens of pathogens to which there is no effective vaccine at present. Subunit vaccines represent a reductionist approach, eliminating the problems with whole organism vaccines. In subunit vaccines, antigens are separated from the infectious element. Synthetic peptides represent a minimalist approach, which could be particularly valuable in examples such as influenza virus where there are constant changes in the antigenicity of the haemaglutinin molecule. Another much longer term approach is to genetically engineer vaccinia (the original smallpox vaccine virus) to carry surface proteins from other viruses, thus making vaccinia a eucaryotic vector for the expression of viral antigens (Wilson, 1984). This will take years of testing and will encounter opposition to the re-introduction of this virus at all, but it does serve to illustrate the exciting range of possibilities now opening up.

Results for synthetic peptides so far have been poor and even results

Table 11.3. *US market for major recombinant vaccines*

Initial market for new vaccines (total for first two or three years)

Vaccine	Immunizations (millions of people)	Price, $ (per immunization)	Values of sales, $ (manufacturer's level)
AIDS	40	20	800 000 000
Herpes	20	20	400 000 000
Total			1 200 000 000

Annual market (once recombinant vaccines are established)

Vaccine	Immunizations (millions of people)	Price, $ (per immunization)	Values of sales, $ (manufacturer's level)
AIDS	2	15	30 000 000
Herpes	1	15	15 000 000
Hepatitis B	3.5	10	35 000 000
Influenza	3.5	10	35 000 000
Pertussis	3.5	10	35 000 000
Polio	3.5	10	35 000 000
Total			185 000 000

From Anon. (1984). US market for human recombinant DNA vaccines. *Genetic Technology News*, **4**, (6), 6–7.

with viral subunits have often been disappointing, with transient and weak immune responses. Fortunately there are exceptions, most notably with the gene coding for a hepatitis B virus surface antigen. Hepatitis B is a world-wide problem, infecting an estimated 200 million people and associated with a high incidence of cancer and other liver disease. It is readily transmitted and can infect the unborn child. The available hepatitis B vaccine (Merck) is expensive (over $100 for a three dose regimen), but even so over a million doses have been distributed since 1982. There are also risks associated with this whole vaccine, because it is produced from the blood of hepatitis B carriers who may also carry other diseases. There are arguments that the price sensitivity of such a product is low, but this is questionable because 80% of hepatitis B carriers live in Asia and Africa and the cost of a mass immunization programme is a prime obstacle. Clinical trials on the recombinant vaccine (Biogen) have begun, with its cost being put in the $10–20 range per treatment. Some authorities argue that the price must be lower still for widespread use in Africa and Asia. A whole vaccine production plant is also being set up in Singapore where 10% of the population are hepatitis B carriers.

The overall potential markets for vaccines are as difficult to appraise as those for interferons or lymphokines. The current vaccine market is small because the products are mostly old and their prices low. The big opportunities are for diseases such as AIDS or herpes where neither a vaccine nor cure is available. A large proportion of the population might be immunized with them during the first two or three years on the market, but after that sales would drop. In a recent forecast the initial markets for these vaccines were put at over a billion dollars for the first two or three years with annual markets for all recombinant vaccines settling at around $200 million (Table 11.3). This type of forecast, while subject to all kinds of reservations, does illustrate the reasoning which underpins the expansion of biotechnology companies and the scale of benefits they can confer.

Other health-care products

Recombinant DNA technology permits the synthesis of many protein products of medical value such as factor VIII for haemophiliacs, urokinase and plasminogen activator for dispersal of blood clots, human serum albumin for fluid loss and other products with considerable market potential.

Veterinary medicine

Most of the biotechnological products for treatment of human disease have their counterparts for animal disease and in some cases market sizes may be equivalent or even greater. Less stringent regulatory procedures ensure more rapid commercialization, but product sales are much more cost-sensitive and marketing may present problems such as convincing farmers of the efficacy of a product and operating through the agricultural supply industry.

Growth hormones and vaccines figure largely in the first recombinant DNA products aimed at the animal market (Table 11.4). Monoclonal antibody kits may be used in veterinary diagnosis and monoclonal therapeutics may be developed for some diseases. Growth hormones are claimed to increase growth rates and the efficiency of feed conversion into body weight or protein. Their market potential has been put in the hundreds of millions of dollars bracket although they are still under trial. Vaccines for scours, a neonatal bacterial diarrhoea of cattle and pigs, have led the field. This disease has been estimated to cause the deaths of up to 5% of the 53 million calves in the United States. It is also a major economic problem in pig breeding with a vaccine market estimated in the $100 million range. Recombinant DNA products however face competition from conventional vaccines for both animals. Similarly foot and mouth disease, an economically damaging disease of cloven-hoofed animals, also has traditional inactivated vaccines. It has

Table 11.4. *Biotechnology products for veterinary medicine*

Chicken growth hormone	Foot and mouth vaccine
Bovine growth hormone	Newcastle disease vaccine
Swine growth hormone	Rabies vaccine
Scours vaccines	Monoclonal diagnostics and therapeutics

virtually been eliminated in much of Western Europe by policies of vaccination, and slaughter of infected herds, but is still endemic in South America, much of Asia and Africa. Recombinant DNA vaccines are being developed because the virus has several distinct immunological forms and there have been examples of residual virulence in the present vaccines causing outbreaks of the disease. Again the market for a good vaccine has been put in the hundreds of million dollars range, but these estimates have been challenged as being excessive. It is however probably the largest veterinary vaccine market.

Rabies vaccine is also a candidate for genetic engineering because of the need to produce large quantities of vaccine safely and cheaply for mass immunization programmes in many parts of the world. The virus is found in a single form with known antigens and has attracted the attention of a number of companies. Vaccines for many other animal diseases are being developed with varying estimates of probability of success. Recently there have also been reports of monoclonal antibodies being used to treat scours in cattle at a cost of around $10 per treatment (Cane, 1983).

Fermentation and the chemical industry

The introduction of a genetically engineered organism into a fermentation system can improve cell productivity by increasing the expression of genes and by increasing their copy number. The system's efficiency is improved in terms of productivity, reaction rates and substrate conversion efficiencies, thereby reducing unit costs and capital investment for smaller plant. In addition genes may be transferred into a host organism which is more suited to the process and is economically more efficient. Factors here include thermotolerance, product resistance, flocculation, non-pathogenicity, tolerance of extremes of pH and therefore non-asepsis, lower production of wasteful by-products, low oxygen tensions and others. Similar considerations apply to immobilized cell systems.

Exciting though all these possibilities are, many of the goals have in fact been achieved by conventional selection techniques, some using genetic techniques, others by empiricism. Some fermentation processes are much less sensitive to improvement by genetic engineering than others. The analysis as described in Chapter 3 can be applied in any individual case to determine the cost-sensitive parameters. For example, as described in Chapter 8, a study on the yeast–ethanol fermentation concluded that genetic engineering could only have a very limited impact on process economics because of the overriding factor of substrate cost and the near theoretical maximum conversion efficiencies. Only limited cost reductions from improvements to ethanol tolerance and thermophilic operation were calculated. It must be stressed that even marginal savings in costs such as these can become of significance to the operators of large plants. It is true however that they will not transform the economics of the process.

Many of the bulk fermentation products listed in Table 5.12 are similarly only poorly sensitive to genetic engineering improvements. Xanthan fermentations, for example, already have high conversion efficiencies from glucose, the major costs being incurred in oxygen transfer and downstream separation of an extracellular product. Similarly the organic and cheaper amino acids are already produced in high yields and at good conversion efficiencies. Other amino acid fermentations can benefit however. Their applications are limited by fermentation costs which have not been reduced sufficiently by conventional technology. Tryptophan for example is an essential amino acid with low concentrations in many foods and feeds. In common with lysine and methionine it could have much greater volume sales in feeds, if the price could be reduced below $10 kg^{-1}. So far with conventional fermentation the price has remained high because of low yields, conversion efficiencies and broth concentrations. Phenylalanine, formerly with only limited therapeutic applications, now has bulk tonnage markets in the synthesis of Aspartame, where it is the most expensive component. Several biotechnology companies are working on cost-effective synthesis of both amino acids and there are competing enzyme and fermentation routes. Genex Corporation predict that the present price of phenylalanine at $18–25 lb^{-1} will fall to $5 lb^{-1} in the next few years. Sales of Aspartame, already at $225 million (1983), will benefit from these cost reductions.

Some vitamins and antibiotics have high costs because of poor conversion efficiencies from raw materials. A high proportion of costs goes into producing biomass which only has a very low by-product value. Unfortunately no details of the application of genetically engineered strains to such a product are available as yet, but a set of hypothetical costings for an antibiotic is shown in Table 11.5. Unit costs are halved by the use of a strain which could achieve a broth concentration of 6.0% compared to a conventional strain producing 1.2%. Not only are raw material costs reduced, but there is a follow-on effect on labour, utilities and capital costs, because the higher broth concentrations permit lower recovery costs and smaller equipment and

Table 11.5. *Process economics – conventional* versus *novel production of antibiotics*

	Production costs ($ millions)	
	Conventional: 1.2% final concentration	Novel: 6.0% final concentration
Raw materials	30.2	14.6
Labour	7.3	5.6
Utilities	16.1	4.2
Equipment	19.0	7.8
Buildings	1.6	0.7
Direct costs	74.2	32.9
Overheads (60% of labour)	4.4	3.3
Total costs	78.6	36.2
Annual production (lb)	5 900 000	5 900 000
Unit costs ($ lb^{-1})	13.32	6.14

Data from: J. Leslie Glick, Genex Corporation.

buildings to handle the same volume of product.

Progress has been slow in this area because the pathways to these products are complex and controlled by many genes with diverse chromosomal locations and regulatory systems. In many cases the full biosynthetic pathway of the product has never been elucidated. The potential cost savings differ. They are very high in vitamin B_{12} and some of the expensive antibiotics, but low in the bulk products such as penicillin where conventional methodology has already achieved impressive results (Fig. 5.1).

Recombinant DNA technology can also be applied to produce extracellular rather than intracellular products. For example the attachment of a leader or signal sequence of amino acids to the N-terminal of a protein can result in it being recognized by the cell membrane and exported. Extracellular products are generally cheaper because fewer steps are required for their isolation (Fig. 11.4) and purification is simpler because there are fewer extracellular proteins. As a consequence recovery efficiencies are higher for extracellular rather than intracellular production. Proteins produced intracellularly by high productivity strains have also often accumulated in insoluble, inactive protein granules. In addition the production per unit volume of growth medium is related to cell density for intracellular products. To grow

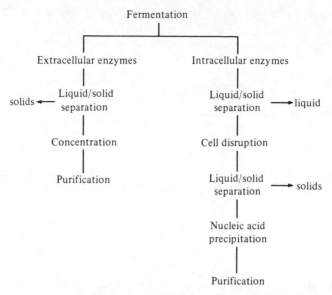

Fig. 11.4. Processing advantages of extracellular enzymes.

cells to a high density requires more raw materials, the efficiency of raw material conversion is lower and often oxygen transfer costs are high. An efficient protein-exporting cell can be grown to lower densities for the same product concentration and it has higher raw material conversion ratios. Again all these improvements will result in lower raw material, labour, utility and capital costs. They explain why biotechnology companies have invested much effort in working with protein-exporting species of bacteria such as *Bacillus*.

In practice problems have been encountered because of protease degradation of extracellular proteins. There are other more specific examples of where genetic engineering can be used to improve fermentation processes. ICI substituted glutamate dehydrogenase for the glutamine nitrogen-fixation system of their SCP organism *Methylophilus methylotrophus*. This system is more energy efficient and has improved biomass conversion efficiencies. It was not adopted however because the additional regulatory requirements (it has to be treated as a new strain) were not considered to be justified. This will continue to be a big obstacle under present legislation when genetic engineering makes a marginal improvement in process efficiency.

Plant biotechnology

Commercial development of plant biotechnology has lagged behind the animal and microbial advances because of a lack of basic scientific knowledge, particularly at the molecular level, and the technical obstacles in the growth and cloning of plant cells. There are now developments in the field which are starting to close the gap. The lack of progress is certainly not because of poor commercial potential. Indeed there are many who would argue, with some justification, that developments in plants could have a greater impact on human welfare than any other area of biotechnology. Advances in conventional plant genetics – selective breeding and cross-fertilization – in the last few decades have brought about big increases in crop yields. They have almost certainly prevented widespread hunger and famine. The resulting fall in the prices of agricultural products, particularly cereals, relative to other commodities (notably oil) has presented biotechnology with its first footholds in the bulk food and chemical markets.

Increases in productivity have not just been brought about by plant breeding. Chemical pesticides, herbicides and fertilizers have had an even greater impact. The side effects of pesticides and herbicides plus the energy requirements or finite availability of fertilizers present biotechnology with other challenges and opportunities.

As a consequence of the infancy of the field even less economic analysis is possible than in microbial or animal genetic engineering. It is worth however entering into a brief qualitative description of work in this area to gain a feel for its potential and for the practical obstacles involved.

Plant cell culture

The growth of plant cells on a laboratory scale makes possible the accelerated propagation of known varieties of food and ornamental plants. Tissue culture and clonal propagation can be used to select new strains via somaclonal and gametoclonal variation and after protoplast fusion. Plant cell culture can also be used for the introduction of new genes via plasmids, viruses and transposable elements. Fermentation is however much more expensive than for bacteria or yeasts because growth rates are slower and growth media are in order of magnitude more expensive. Plant cell walls are fragile and can only tolerate lower shear agitation systems such as airlift fermenters. As with microbes, cell immobilization promises many cost saving advantages. Often the production of useful compounds (mostly secondary metabolites) is restricted to a specific stage in a cell's life cycle and products are excreted into vacuoles rather than into the medium. These factors mean that only synthesis of expensive compounds such as drugs or fragrances can be contemplated at present (Table 11.6). A common estimate is that the wholesale selling price of a product must be at least $100 kg^{-1} for plant tissue culture to be a viable proposition at present. The first compound produced by plant tissue culture, shikonin (a dye and pharmaceutical), is being marketed by Mitsui Petrochemicals Ltd

Table 11.6. Markets for some plant secondary metabolites

Compound	Use	Wholesale price ($/kg)	Estimated retail market ($ million)
Vinblastine/vincristine	Leukaemia	5000 (g)	18–20 (US)
Ajmalicine	Circulatory problems	1500	5.25 (world)
Digitalis	Heart disorders	3000	20–55 (US)
Quinine	Malaria: flavour	100	5–10 (US)
Codeine	Sedative	650	50 (US)
Jasmine	Fragrance	5000	0.5 (world)
Pyrethrins	Insecticide	300	20 (world)
Spearmint	Flavour; fragrance	30	85–90 (world)

From: Curtin (1983).

(Tokyo). It is selling for $4000 per kg (Curtin, 1983). The present consumption in Japan is about 150 kg per year but Mitsui hope to increase markets by price reduction. Their facility can now produce 65 kg per year.

It must also be remembered that tissue culture must have a clear-cut advantage over the normal agricultural or horticultural method which includes extraction from the whole plant. The stimulus for this development is often uncertainty of supply. Many plant secondary metabolites come from tropical countries and have tenuous supply lines. Diseases, adverse weather and political instability have often created shortages and widely fluctuating prices. Many companies will pay a premium to be able to regulate their own production to supply needs.

The alternatives are chemical synthesis, which is long and complex, or bacterial fermentation, generally impossible because of multi-step pathways under the control of unknown genes. Plant cell tissue culture will encounter a problem with such low volume, high price products in that any oversupply in such a limited market will send prices tumbling, perhaps below the economic thresholds of the process.

Plant genetics

Clonal variation and even protoplast fusion are regarded as having only limited value compared to *in vitro* recombination technology in plants. Development is limited, in part at least, by the lack of available vectors. The most commonly used vector, the Ti plasmid from *Agrobacterium tumefaciens*, has a wide host range, but only in dicotyledons. It cannot be used, as yet, for the major monocotyledon cereal crops such as maize. There are reports of plant virus and transposon vectors, but none has gained wide acceptance.

Table 11.7. *Main targets of plant genetics*

Selection of high yielding varieties
Improvement of nutrient efficiency, reduced fertilizer consumption
Introduction of herbicide resistance
Increased disease or pest resistance
Improved quality of food product
Increased ability to grow in poor conditions on marginal land

Microinjection techniques into cell nuclei must still be attempted in many cases.

There is still widespread interest in plant genetic technology and a number of biotechnology companies have been established to concentrate in this area. There is a stronger element of market pull because of the potential for crop improvement (Table 11.7). Estimates are that plant genetics could add $5 billion per year to the value of major crops over the next decade and even more subsequently. The market for seed corn alone in the United States is $1.3 billion annually. Biotechnology companies involved in research in this area tend to collaborate closely with companies engaged in developing and commercializing high profit proprietary plant varieties and with herbicide manufacturers. The most widely recognized goal, that of introducing nitrogen fixation to major cereal crops, is realized to be a difficult and longer term objective. Similarly the lack of knowledge of plant gene regulation has hindered direct development of improvements in productivity. Instead efforts have been directed into herbicide resistance and the introduction of new proteins into crops; in these areas some technical advances have been made.

The herbicide industry is a high added value industry with a US market of $2.4 billion per year. The introduction of plants resistant to a specific compound would permit the selling of a herbicide plus resistant plant package which can extend the life of a profitable product. For example Calgene have been able to isolate bacteria resistant to a glyphosate herbicide and aim to transfer the gene conferring resistance into crop plants such as tobacco, soya and cotton. Glyphosate is the major ingredient of 'Roundup', the world's largest selling herbicide now marketed by Monsanto, but reaching the end of its patent life. It earns around $400 million per year and with one other herbicide accounts for 13% of Monsanto's sales (Yachinski, 1983). The sale of these plants would aid sales of glyphosate, which is effective at low application rates and disappears quickly from the soil, but is non-selective – it damages crop plants as well as weeds. Similarly Ciba-Geigy is reported to be developing soy beans resistant to the herbicide atrazine, which no longer has patent protection, but would permit crop rotations of soy bean after corn treated with atrazine.

Another area of plant genetics which in fact could displace other biotechnological products is in the improvement of nutritional quality. Corn in particular is deficient in some essential amino acids such as lysine and methionine. These amino acids are often added as supplements in normal feeds. If crops could be produced with proteins rich in these amino acids the supplements could be eliminated. Recently the gene for phaseolin, the major storage protein in bean plants, has been transferred to tobacco, where it has been expressed (Netzer, 1984). Commercial applications are still some way ahead here, as elsewhere in plant genetics, but the wide-ranging potential benefits have succeeded in attracting appreciable capital investment.

Microbial pesticides

The use of microbial pathogens of insects as a form of biological pest control is an elegant concept which has been studied for some time, particularly with the rising incidence of resistance to chemical pesticides and their toxic effects on the environment. It has however met with only very minor success in the field. The chemical business world-wide is worth some $13 billion annually in gross sales, but microbial pesticides amount at the most to some $20 million in sales. Of over a thousand known microbial pathogens of insects only a handful are approved for commercial use, despite the fact that they are specific and non-toxic to animals.

The only widely used bacterial insecticide has been *Bacillus thuringiensis*. It is active against many lepidopteran larvae and is mostly used on pests of brassicae (the cabbage family) and cotton. In the 1960s there was widespread commercial interest in *B. thuringiensis* and some 20 small companies were involved in its manufacture, using conventional batch fermentation, drying of crude broths and formulation for spraying. Today the bulk of the *B. thuringiensis* market is supplied by three manufacturers (Abbott, Sandoz and Solvay). The total market is worth some $10 million annually and the product is only considered viable if produced in a multi-product facility. The market has never been large or lucrative enough to justify the construction of a new plant dedicated to *B. thuringiensis* production. In recent years interest in this organism has been rekindled by the discovery of the *israelensis* strain which is active against mosquito larvae, including malaria-carrying species. Some other *Bacillus* strains, notably *B. popillae* and *B. lentimorbus*, are also made in limited quantities commercially.

There are a number of reasons for the poor market penetration of

Table 11.8. *Microbial* versus *chemical pesticides*

Advantages	Disadvantages
Selectivity which is particularly important in environmentally sensitive markets such as forestry	Lack of reliability or slow speed of kill in less sensitive pests
Can be used immediately before harvesting which can be important for vegetables	Environmental factors such as humidity may affect efficacy
Strong theoretical background	Lack of persistence or short field life

biological control agents. One is price. Estimates on the cost of use of *B. thuringiensis* range from 20–25% more expensive than comparable chemicals (Klausner, 1984) to up to 10 times more expensive to spray than DDT or other chemical pesticides used against mosquitoes (Lisansky, 1984). The current production cost of *B. thuringiensis* is difficult to establish, but is probably around $8–10 per kg dry weight of cells of which 20% is the insecticidal toxin. The formulated products retail for $10–20 per kg and are usually 10–20% cells.

Some advantages and disadvantages of microbial pesticides are listed in Table 11.8. The microbial products are very species-specific, they often only fit into sophisticated pest management schemes and frequently will not work unless strict methods of application are followed. On the other hand there is room for much long term optimism if genetic techniques can be applied to strain improvement. Recombinant DNA technology could be used to increase virulence (thereby reducing manufacturing costs for the same insecticidal effect), and to increase host range, permitting simultaneous protection against a range of pests and increasing field life through resistance to adverse conditions. Some support for this approach has come through the discovery that the insecticidal toxins produced by *B. thuringiensis* strains are plasmid encoded and may be transferred to other bacteria. Insect viruses have received far less attention so far than bacteria, but many experts in the area predict that greater levels of virulence will be attainable with them. Pathogenic fungi have the attraction that they can infect insects by contact; they do not have to be ingested. Another possibility is to clone toxin genes directly into plants.

New engineered organisms require expensive regulatory approval, unlike wild isolates which are subject to easier registration conditions and are cheaper to introduce than a chemical pesticide. Overall, the area is very costly in terms of treating, regulation, fermentation, downstream processing formulation and marketing. It has very high entry barriers. New independent companies will find it very difficult to compete with established pesticide manufacturers. Many biotechnology companies are involved however, often as partners or performing contract research for major agricultural chemical concerns.

Summary

The development of new biotechnology has been shaped by technological advances. It has given rise to 200 new companies in a few years and attracted over $2 billion in investment, mostly in the United States. No other industry has been launched with such vigour before real products were clearly identified. The enthusiasm is due to the medical subjects involved and the present cult of high technology. There are impressive reports about potential benefits and predictions of huge markets, but there is a danger of short term expectations being too high. When microelectronics firms launch a new model they can usually expect to see some return on it within a year. In biotechnology large amounts of cash and many years of development time are required. As a rough approximation it has been estimated that it costs $1 million to clone a gene if nothing goes wrong, and up to $2 million if it does – and this is only the start (Hester, 1983). The product must be produced by fermentation using the right strain and at high efficiencies, purified, tested, given toxicological approval and so on. Even pilot plant facilities start at $20 million. Given all these factors biotechnology has been treated almost uniquely by investment analysts and the stock market. Widely predicted shake-outs have not occurred, companies having continued to attract funds, often by means of new forms of financing such as research and development limited partnerships. Nevertheless the industry still faces two problems: of over-concentration in the health-care field, and cash flow to sustain expenditure on new products before earnings build up.

The concentration on health-care is the consequence of a number of factors. First, products are high value and require small production facilities. Second, the technology leans towards single gene products such as interferons and peptide hormones. The wider applications in fermentation chemicals and plants are usually dependent on multi-gene, multi-operon pathways which present greater technical obstacles. The historical success of antibiotics has also probably played an important role. The cash flow problems may be overcome by further investment schemes, greater participation or takeovers by large companies, or by good commercial successes in the first generation of products.

In contrast to the economic uncertainty it must be stressed that the pace of recombinant DNA research has in reality been much faster than even optimistic forecasts predicted a few years ago. In the late 1970s it was predicted that expression of insulin by recombinant DNA organisms was five years away. In reality it was achieved in under two years. Similarly expression of interferon, predicted to be a decade away, was achieved in two to three years. Automated DNA synthesis can make oligonucleotide sequences in a tiny fraction of the time taken by the manual phosphotriester method, which itself was 10 times faster than the older diester method. DNA sequencing too can be achieved in a fraction of time and at a fraction of the cost of a few years ago. Genetic engineering may in fact turn out to be more valuable as a tool for understanding biological processes than as a means of production. It could also lead to more efficient and precise drug development from which the pharmaceutical companies, with their present expertise in manufacturing, have the most to gain. It may take a combination of techniques and new advances before commercial success is widespread.

Further reading

The Gene Age. (1983). E. J. Sylvester & L. C. Klotz. New York: Charles Scribners & Sons.

New developments in this area are covered by numerous journals and news sheets such as *Biotechnology*, *Genetic Technology News*, *Biotechnology News* and *Practical Biotechnology*.

12 Prospects for the economic development of biotechnology

Historical perspective

In comparison to the spectacular advances and discoveries made in the understanding of biology, the fortunes of biology's commercial application have been mixed. Successes such as antibiotics and amino acids have been countered by failures such as single cell protein or losses to the chemical industry with respect to solvents or some of the vitamins. The speed of developments compared to other technologies has sometimes been slow. For example *B. thuringiensis* was first isolated by Ishiwata in 1902 from dying silkworm larvae. It was first grown as a pesticide in France in 1938, but today the impact of microbiology on the world pesticide market is still small. Similarly the first report of an immobilized enzyme was in 1908, the acetone–butanol fermentation was developed during the First World War, and commercial ethanol plants pre-date this. Anaerobic digestion is centuries old. Thus progress has been very slow compared to other technologies such as electronics or aviation over a similar time period.

There have nevertheless been significant advances in biotechnological processes which have permitted cost reduction and increased sales volumes of many products. In an interesting review on the development of the fermentation industries since 1910, Perlman (1978) examined the evolutionary patterns and proposed three major sources of stimulation of innovation:

(1)   The occurrence of crises and their resolution
(2)   The availability of new technology
(3)   Interference from outside influences, notably politics.

Crises include scientific problems and those of an economic origin such as raw material price rises. Scientific problems of note include bacteriophage infection, allergic reactions to enzyme washing powders, and environmental pollution. The political crises include the need for acetone in the First World War, the oil price rises of 1973 and 1979 and the foreign debt of Brazil. New technology includes both hardware in the form of reactors and plant design plus 'software' in selection or construction of new microbial strains. Some notable achievements include the development of immobilized enzyme reactors, membrane

276

technology, high antibiotic- and amino acid-producing strains, advances in immunology such as the discovery of interferons and lymphokines and now strains yielding products coded for by recombinant DNA. Political intervention has taken the form of anti-pollution legislation, legislation on the testing of food and drugs, subsidies for fuel production and grants or tax relief for the construction of new productive facilities. A significant fourth factor which should perhaps be included is the availability of raw materials, manifested as the relative change in the price of oil and agricultural feedstocks, although it can also be argued that the availability of surplus grain is largely a political and technological phenomenon.

The present position of biotechnology in national economies

The result of the above-mentioned influences has been to create an industry which in economic terms covers a diverse range of products linked by a common means of production. Biotechnology tends to fit certain niches, often in food and drug production, which cannot be provided by chemistry or agriculture. In total sales it is dominated by ethanol, antibiotics and high fructose syrups with significant contributions from amino acids, organic acids and enzymes. New ventures using recombinant DNA technology are in the limelight. They have attracted prodigious volumes of capital backing and have high research expenditures, but it is too early to measure the impact of their products. Biotechnology is still small in relation to the oil and chemical industries, but its largest selling products are entering the same sales ranges.

Biotechnological products can be conveniently, though arbitrarily, assigned to low, medium and high price categories. The low price and high volume category can in general be identified as low molecular weight sugars, alcohols, aldehydes, ketones and acids used as cheap food ingredients, feedstocks, fuels and solvents. Their production costs are most sensitive to the cost of raw materials and their production by biotechnology is in general still not feasible without political intervention. Those products competing with petrochemicals will not become viable without further oil price rises and the dependence on raw materials means that a large quantity of feedstock must be available or shortages will cause adjustment to a new price equilibrium.

The middle price-range products include food additives and speciality chemicals. Only a limited number are produced commercially, such as amino acids, nucleotides and xanthan gum. This area may be described as applications limited because few of the numerous compounds occurring in metabolic pathways or as extracellular products have been found to have commercial applications so far. Many microbial products have potentially valuable functional properties such as surface activity, chelation or complexing properties, but in general they are more expensive than their chemical counterparts. Xanthan gum is a notable exception. It is used in oil drilling muds and as an emulsifier, stabilizer and thickener in many industrial and food applications. Some microbial products with interesting functional properties, such as cyclodextrins and many polysaccharides, are still short of applications. In some instances the chemical industry may be able to use biological compounds as models and improve upon them by chemical synthesis.

The high value product category, already well represented by antibiotics and vitamins, is the main target of genetic engineering venture companies. Their first generation products are mostly proteins with therapeutic and veterinary applications. They have been discovered as a result of medical research and not by the predominantly empirical screening of the pharmaceutical industries; if successful they would therefore represent an advance for rational chemotherapy. If their promise holds up, the biotechnology industry will have an assured future; if not, many companies will experience severe financial difficulties. Enzyme technology may also be described as applications limited. Enzyme penetration of starch and detergent markets is approaching saturation and although the small scale diagnostic market is healthy and growing there is a need for new bulk enzyme applications to sustain the growth experienced during the 1970s and early 1980s.

There are also other applications of biotechnology, such as pesticides, mineral leaching, concentration of metals, enhanced oil recovery and even electronics, which do not conveniently fit under the three headings system at present. Most are still in development stages with little commercial impact and futures that can only be speculated on. Mineral leaching is an exception. The leaching of copper from low grade sulphide ore dumps by *Thiobacillus thiooxidans* accounts for more than 10% of US copper production. It is still essentially an empirical process, often with low extraction efficiencies, but could be inexpensive and has low energy requirements. If optimized it could replace the smelting of ores with the latter's high capital and energy costs and pollution. At the current low price levels of bulk metals the best opportunities appear to be in high value metals such as uranium. One other application which merits reference is the use of biosensors for the measurement and analysis of a wide range of products. A biological sensing element, often an enzyme, immobilized on a membrane is integrated to a transducer which produces a quantifiable electronic signal. These sensors can be used to detect and measure biological compounds, microbial contamination or biomass directly; i.e. they are on-line measurements. They are rapid, sensitive, specific, use small sample volumes, can be miniaturized and potentially could become very

Table 12.1. *Strengths and weaknesses of biotechnological processes*

Strengths	Weaknesses
Reliance on renewable feedstocks	Feedstocks are oxidized and unsuitable for many applications. Their costs also fluctuate
Low temperatures	Sterilization a major cost
Operation in aqueous media	De-watering of dilute solutions a high cost
Several reactions may be covered in one fermentation step	Downstream processing expensive
Plants can be highly automated	High capital cost of equipment
Stereospecificity	Reactions often slow – poor volumetric productivity
Less toxic chemicals for disposal	High BOD wastes

cheap. Their use in the measurement of blood components and metabolities could revolutionize the diagnosis of disease. They can also monitor pollution, be used in agriculture and in the control of biotechnological process industries. This is an area which could see rapid commercial growth in the near future, probably by small, innovative firms.

Process factors

In view of the sometimes slow progress of biotechnology and the limitations of its present day range of applications, it is worth reviewing the strengths and weaknesses of the technology (Table 12.1). The weaknesses are all particularly telling at the low price, high volume end of the market where there is direct competition with other technologies. The products which do succeed in this category all manage to overcome at least some of these drawbacks even though they are still generally protected by legislation. Thus the ethanol fermentation can be run at low pH without the need for sterilization, it uses cheap feedstocks, simple equipment and the product can be recovered from broths by distillation, a relatively inexpensive single stage operation. The production of glucose and high fructose syrups is only possible because of cheap substrate (corn starch), cheap saccharifying enzymes, immobilization of glucose isomerase and operation of relatively high temperature and high concentration streams to increase reaction rates, reduce reactor costs and cut contamination problems. Single cell protein, on the other hand, fell foul of all the disadvantages and if it is to succeed must overcome at least some, for example cheap feedstock and reduced downstream processing costs.

It must also be stressed that in common with other new technologies research in biotechnology is expensive and often an appreciable cost factor in many projects. Cloning a gene has been costed at a minimum of $1 million, and more if any difficulties arise. Novo have estimated that at least 20 man-years have to be devoted to the isolation of a commercially useful new enzyme. A useful antibiotic is obtained for every 20000 strains of organism screened.

Biotechnology in relation to theories on technological development

An introduction to a recent biotechnology conference proclaimed that as surely as physics and chemistry dominated the twentieth century, so biology will dominate the twenty-first century. This is a widely held belief, but if it is to be realized many improvements are called for. It is worth examining what has to be achieved and how it might occur, by reference to accepted economic models of technological development.

Technologies have come to be viewed in terms of a Schumpeterian discontinuity followed by refinements that determine the rate of productivity growth to reach a peak (Fig. 3.1). As described earlier, this model has some limitations when applied to a whole technology because some sectors may be in different phases of development. Thus in biotechnology many fermentation techniques are in a mature phase, while much genetics can be regarded as being in an innovative phase. The whole technology curve must be regarded as the sum of a number of innovations and developments, all of them discontinuous, but summing to a smooth curve. If it is assumed that the majority of developments in biotechnology are still in the innovative phase, the steeply ascending growth phase will only be achieved through the interaction of a number of phenomena. The factors which go together to contribute to the development of a technology have been described most thoroughly by Rosenberg (1982a). The use of his models can provide a valuable base to explore the future pattern of events in biotechnology.

The enabling phase may be viewed as the type of development work that is going on in genetics and molecular biology. For example genetic engineering has enabled species barriers to be overcome; it has permitted the expression of a gene from one organism in another, which is enabling the bulk production of formerly rare and expensive proteins in a commercially suitable organism. Another potentially valuable enabling technology is found in enzyme engineering. The detailed study of enzyme structure can show which amino acid residues and which sequences confer structural stability and catalytic activity. The study of enzymes which operate at extremes of temperature or pH can provide further information. Combining this knowledge with techniques in molecular modelling may make it possible to design new enzymes which have combinations of desirable characteristics such as thermostability, high catalytic activities or new catalytic activities and an absence of side reactions. Using present technology the appropriate base sequences can be constructed, inserted into DNA, and high levels of expression obtained in a suitable organism.

The enabling phase must be followed by detailed developments which address themselves to the limitations of biotechnology listed in Table 12.1. These will include reactor design, membrane technology, and adoption of new methods of de-watering and downstream recovery. To a large extent progress in this phase will be dependent on empirical testing and learning by use. The refinements are often small – a succession of minor modifications. The result is the real improvements in productivity and cost reduction which permit the acquisition of large markets.

The rate of adoption of the technology depends on the assessment of risk by entrepreneurs and corporate decision takers. Some of the high risk associated with biotechnology is now being reduced by government policies, subsidies, grants, capital allowances, the US capital gains tax and so on. There is also the factor of timing. Often trail-blazers go bankrupt and those who later buy the equipment, buildings or the expertise make money out of it. This is a factor of importance to genetic engineering groups in particular: long lead times and regulation mean that large financial commitments are required when the risk is highest, so time is crucial. On the other hand with correct timing the first innovator can experience abnormal profits until his patent expires or the others catch up with him. The refining and development of a process tends to lead to an increase in the size and concentration of firms through economies of scale and ultimately to an oligopolistic industry structure.

Technological interdependence

Inventions or technologies do not function in isolation, they are dependent on the availability of complementary technologies. There is a process of cross-fertilization and mutual reinforcement which can be seen in historical developments such as coal, steel and railways or petroleum and chemicals. The interaction of many technologies can be seen clearly in the corn wet milling industry, which is dependent upon advances in agricultural technology for cheap grain, in biotechnology for cheap, thermostable enzymes and in engineering and process design. One invention can increase the economic benefits of another. Biotechnology in general can benefit from inventions and innovations which permit the handling of aqueous solutions, improve separation of solutes in heterogeneous mixtures, reduce reactor costs, save energy and so on. Often there is some delay between the development of new products and their commercial exploitation. For example the discovery of many of the major polymers such as PVC, polystyrene and polyethylene in the 1930s and 1940s was not exploited until the 1950s. It was aided by the drop in the cost of petroleum feedstocks. Similarly biotechnology may be dependent upon further falls in agricultural feedstock prices. The benefits of a new technology filter down: the chemical innovations in polymers led to innovations in textiles later. The availability of cheap biotechnological products may stimulate new applications, for example in compounds such as methyl glucoside which is now available at a lower price from wet milling operations.

Biotechnology is also benefiting from advances in electronics. Computers improve process control and efficiency. Biosensors are an alliance of enzyme and semiconductor technology. The close coupling of redox enzymes to electrodes may also make possible the synthesis of new speciality chemicals.

Threshold levels of economic viability

There is a threshold level at which the costs of new technology become competitive with those of the old. Ocean-going passenger liners could still compete with piston-engined aircraft, but once jets appeared the battle was lost. There may be a long gestation period in the development of new technology during which gradual improvements are not exploited because overall costs are still in excess of the old (Rosenberg, 1982b). As the threshold is approached and eventually crossed the adoption rates of the new technology may become increasingly sensitive to further improvements. Large technical improvements may be made in the first period without economic or commercial results, while conversely after the threshold has been crossed even small improvements may lead to important productivity gains. The threshold is of course dependent on all the costs, including raw materials. The principle may be illustrated by reference to fermentation ethanol. Corn costs and the subsidies which have led to the creation of the wet milling industry have led to the production of fermentation ethanol which is on the threshold; it can be produced for approximately the same price as ethanol from petroleum naphtha. It is however still below the threshold as a fuel in competition with petroleum, unless this is artificially altered by taxation policies. Further increases in oil prices will tip the balance in favour of the fermentation product.

Since biotechnology can make many products in competition with

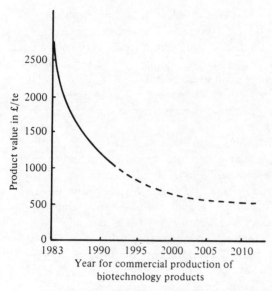

Fig. 12.1. Predictive timetable for commercial production of biotechnology products. From M. Stewart (1984). The role of the biochemical engineer. In *The Proceedings of Biotech. '84 Europe*, pp. 247–63. Pinner, UK: Online Publications Ltd.

chemistry, the threshold for different products will be crossed at different times. Thus because biological feedstocks are predominantly carbohydrates the threshold may be crossed first for oxidized products such as aldehydes or acids, and only later for alcohols or polymers. Alternatively because biotechnology has high process costs, and cheap products are dependent on cheap raw materials, the threshold may be crossed progressively with decreasing price (Fig. 12.1). The threshold will also be reached in different nations at different times depending on the relative costs of agricultural or petroleum feedstocks. Ultimately the rate of diffusion of biotechnology will be dependent on the speed at which it can confer economic advantages over its competition. This may be a fairly smooth progression world-wide, or more probably it will be a more piecemeal process dependent on specific products and local conditions.

The influence of the market

New technologies do not just reduce the cost of manufacturing existing goods. More significantly they produce new products via innovation. Industrial societies enjoy a higher standard of living today, not only because they produce the goods which pre-date the industrial revolution more cheaply, but also because they manufacture new products. This process of innovation often creates a problem of finding applications and therefore markets for the new products. For example chemical polymers, which now seem indispensible, were once inventions in search of applications. For some years polyethylene only found use as a high voltage insulator. Technology push must be balanced by market pull. There is often an inertia barrier to overcome in inducing potential customers to switch to new products or new ways of doing things. It may take years of persistence to bring new products to the market and to derive earnings from them. It is essential to get feedback from the market place, to identify gaps and therefore opportunities and to exploit them. For example there are a large number of known microbial polysaccharides; they have a great range of rheological properties and many are surface-active. They may be used as thickeners, stabilizers, suspending agents, emulsifiers and demulsifiers, yet so far only one product, xanthan, has achieved any sizeable degree of market success, and then only after many years of development. Similarly microbes produce a wide range of surface-active molecules, from sugar lipids to lipoproteins, few of which, if any, have found commercial use. The microbial polymer poly-β-hydroxybutyrate is also in search of applications. The list can be extended to many products of primary and secondary metabolism.

It is important to make a distinction between the precise concept of market demand and the diffuse one of human needs. Demand as mediated through the market-place is a systematic relationship between prices and quantities (Fig. 2.2). Demand and technological opportunity each represent a sufficient condition for innovation to occur; each is necessary but not sufficient for innovation to result, and both must exist simultaneously (Mowery & Rosenberg, 1982). In biotechnology demand is difficult to describe overall because of differences in market sectors. For example demand for surfactants is specific, complex and requires much interaction between manufacturer and consumer to identify specific applications. In many areas of medicine, however, demand is very clear – to diagnose or treat a given disease. In affluent societies too, high cost medical care is encouraged for the individual, regardless of its cost to the society.

In conclusion there are no hard and fast rules for success. Those manufacturers who select the right targets judiciously and match their innovation to the market-place will succeed. It is essential that they maintain a dialogue with their potential customers.

Biotechnology in relation to long wave cycles in economic development

Since the industrial revolution and the growth of capitalist societies there have been cycles of economic activity which consist of a rapid expansion followed by relative stagnation and depression. These so-called 'long waves' or Kondratiev cycles span approximately 50–60 years. There have been many attempts to explain them; for example the production and duration of long term capital was considered to be the leading and most important determinant (Kondratiev, 1935). Other factors such as labour, raw materials and entrepreneurship have been implicated in lesser ways and the periodicity of cycles is blurred by events such as wars or political upheaval. It has been argued by others, notably Schumpeter (1939), that there are clusterings of basic innovations during the period of depression (for example the 1820s, 1870s and 1930s) and that these innovations generate new sectors of the economy. New markets and the availability of capital in the depression permit these sectors to grow rapidly. This creates an upswing in the economy, generating further new investment and employment on a large scale which is based on the scientific and technical development work of the previous decades. The upswing coincides with the growth phases of the new technologies involved, as shown in Fig. 3.1. The growth phases are followed by market saturation, poorer returns, and less investment. Stagnation and decline then set in. These theories are discussed thoroughly in a collection of papers by leading exponents of the field (Freeman, 1984).

It is tempting to view biotechnology in the light of these theories, with new innovations such as genetic engineering and monoclonal antibodies clustered in a period of economic recession and attracting large amounts of capital investment. It would then be predicted that biotechnology would contribute to the subsequent economic upswing, just as did railways in the 1830s and 1840s, the early electrical industry in the 1880s and 1890s and chemical polymers in the 1950s.

There are however a number of cautionary points. There is not a direct parallel between long wave and technology life cycles. Long waves are also caused by fluctuations in investment in industrial infrastructure and major innovations can extend their potential over more than one expansion phase (for example automobiles). Some important industries, notably computers and microelectronics, have in recent years had an expansion phase through an economic recession. Also in one area, medicine, a major biotechnological sector, there does not seem to be a relationship between major innovations and business or industry life cycles (van Duijn, 1984). Major technologies require not one or two but a number of related inventions, and often organizational changes which together allow them to grow. Biotechnology has already spanned two complete cycles since the first industrial ethanol fermentations in the 1880s. It is now entering a critical phase when the number of successful innovations will determine whether or not a rapid expansion phase ensues as a constituent part of a general economic upswing.

Impact on world trading relationships

A major concern expressed about the growth of large scale biotechnology based on agricultural feedstocks is that it will divert land and crops away from food and could exacerbate the plight of the world's undernourished. In addition, increased demand for commodities will simply push up their price to a level too high to contemplate their use on a large scale. There has been some evidence for this effect when Brazil diverted molasses into ethanol production. The molasses price rose so that it became more profitable to export the molasses directly. Molasses, however, is unusual in that it is a by-product whose supply is limited by sugar-refining and is not influenced directly by demand. In general the prediction of the effect of major programmes on world commodity markets is a complex subject beyond the scope of this work. The reader is referred to specialist articles on this subject, for example Chattin & Doering (1984). There is some evidence that biotechnology will not have this kind of impact in the long term. World grains production is rising, with many countries (notably the United States, Canada, Australia and Argentina) now having large surpluses and even Western Europe moving into surplus. The surpluses will either have to be subsidized by governments or production will have to fall to maintain an equilibrium. The evidence is that governments are favouring diversion to other uses such as biotechnology. Genetic improvement of crop strains will permit even greater production. A recent US Office of Technology Assessment (January 1984) estimated that 14% of the US corn crop (around 270 million te) could serve as chemical feedstocks by 1990 without jeopardizing supplies required by other users. Also less than 10% of US corn production is used for food, over 70% being used for animal feed. Similarly world sugar production (cane and beet) is in excess of that required for human consumption. As a result the market is glutted, prices are at an all time low and many sugar producers can scarcely cover their marginal costs of production. The demand for sugar as a sweetener in industrial nations is predicted to decline slowly. As with grains, either production must fall or new uses must be found. Where the land is required for another crop, as was the case in Kenya, there is a direct conflict with food, but in Brazil this is not the case. Brazil does have a grains shortage, but cereal crops cannot be grown in the sugar cane regions.

Overall it is too early to predict the effects of large scale demand for renewable resources. There may well be conflicts, but there are some grounds for optimism. Demand for crops as feedstocks could work to the benefit of many developing nations who are suffering from the effects of low prices at present. It could also shift world trade balances in their favour in the longer term, giving them potential to develop industries on indigenous crops.

As a caveat it must be stressed that world energy requirements cannot possibly be met by crop production in the foreseeable future. Converting all the corn grain produced in the United States in 1981 (8.1 million bushels) into ethanol would supply 21.1 billion US gallons of liquid fuel or about 14% of US motor fuel use (Wagner *et al.*, 1983). Limited developments such as the supply of some speciality and bulk chemicals could be achieved. Only improvements in cellulose and lignocellulose breakdown can lead to any significant supply of fuel.

Conclusion

In the short term a number of products are being developed by biotechnology companies. Foremost among these are diagnostic kits based on monoclonal antibodies, insulin, new amino acids (notably phenylalanine for aspartame manufacture) and ethanol as a petroleum octane improver. The sales of these products could go some way to establishing confidence in the industry and overcoming the threshold barriers so that detailed improvements in the productivity of the technology can be achieved. In the longer term many more developments are possible which can have a much greater economic impact. Most important perhaps are the genetic improvement of crop plants, production of immunological reagents such as lymphokines and engineering of enzymes for new functions. These and other possible developments are so broad-ranging that it is impossible to perceive their impact. What is important is that biotechnology will not just replace the chemical industry or any other industry directly, it will open up new possibilities and new products as other technologies have done in the past.

In the short term the industry has a number of economic problems to contend with. In common with other industrial technologies of the twentieth century it is increasingly dependent upon complex and expensive scientific research with long lead times on products. There is a high degree of uncertainty in predicting the eventual performance of many of the products. There is also an overconcentration of resources on too few commercial targets which, combined with the research costs, means that there is a danger that no one will receive an adequate return on investment. In addition, new plant is capital intensive, there is a lack of engineering expertise in designing successful plants and a need for wide-ranging process improvements. These will probably come about once sufficient new products reach their economic threshold. This situation would tend to favour the survival of a few leading companies which successfully develop new products early and can corner the market long enough to develop the production resources. Weaker, slower and second rank companies will not have high survival rates.

In many nations there is a strongly held view that the cure for industrial decline and rising unemployment is the development of new firms in high technology areas. It is a view which is often based on uncertain evidence and inadequate understanding of the issues involved. In biotechnology most countries wish to emulate the United States but are uncertain as to which strategies to adopt. The United States is wealthy enough to afford high risk programmes – the 'shotgun approach' – in which some firms succeed and form the large scale industries of the future while the majority fail or are taken over. This approach has been successful in the past with other industries, but elsewhere industries with comparable success have been built on different lines. The Japanese for example seem to be adopting the same approach with biotechnology as they did with automobiles and electronics. While not being first in the field in genetics, their expertise in process development and product innovation may eventually put them in a very strong position.

Other aspects of the American development may be emulated elsewhere. The rapid spontaneous growth of high technology industries has occurred in particular locations such as Massachusetts and California. The availability of capital, although important, is only one part of the explanation. Other factors such as skilled labour, specialist links with academic institutes and technical information sources have undoubtedly been contributory. Science can be linked with the business community through university science parks, joint development programmes, industrial funding of university research, groups which share information and other means. These schemes may take on a crucial role in the development of biotechnology.

At the bulk commodity, low price end of biotechnology, policies are shaped by national resources, notably the availability of petroleum and the capacity to produce agricultural feedstocks. These policies can be criticized on the basis that they divert resources into propping up inefficient industries, or they may force manufacturers to pay more for indigenously produced feedstocks, which may make their products less competitive on world markets. They can, however, reduce foreign exchange losses and create employment. Perhaps more beneficially they can divert agricultural resources away from supplying already glutted world markets with prices below production costs in some instances. With many agricultural products, notably cereals and sugar, producing nations must either divert use into new applications or cut production. This may well be the greatest single influence in the development of large scale biotechnology. If the competitive threshold is lowered in this way, the subsequent process developments (as has already been seen with ethanol distillation in America) can radically improve the economics.

References

Chapter 1

Atkinson, B. & Mavituna, F. (1983). *Biochemical Engineering and Biotechnology Handbook*. London: Macmillan.

Bradbury, F. R. & Dutton, B. G. (1972). Objectives. In *The Chemical Industry: Social and Economic Aspects*, pp. 11–20. London: Butterworths.

Bull, A. T., Holt, G. & Lilly, M. D. (1982). Introduction. In *Biotechnology – International Trends and Perspectives*, pp. 21–38. Paris: OECD.

Fishlock, D. (1982). Pitfalls of biotechnology. In *The Business of Biotechnology*, pp. 69–86. London: Financial Times Business Information Ltd.

Godley, W. & Cripps, F. (1983). Introduction. In *Macroeconomics*, p. 13. Oxford University Press.

Rosenberg, N. (1982). How exogenous is science? In *Inside The Black Box: Technology and Economics*, pp. 141–59. Cambridge University Press.

Chapter 2

Bradbury, F. R. & Dutton, B. G. (1972). Selling chemicals. In *The Chemical Industry: Social and Economic Aspects*, pp. 113–38. London: Butterworths.

Bull, A. T., Holt, G. & Lilly, M. D. (1982). *Biotechnology: International Trends and Perspectives*. Paris: OECD.

Cooper, M. H. (1966). *Prices and Profits in The Pharmaceutial Industry*, p. 274. Oxford: Pergamon Press.

Duncan, W. B. (1982). Lessons from the past, challenge and opportunity. In *The Chemical Industry*, ed. D. H. Sharp & T. F. West, pp. 15–30. Chichester: Ellis Horwood.

Egan, J. W., Higinbotham, H. N. & Watson, J. W. (1982). *Economics of The Pharmaceutical Industry*, p. 205. New York: Praeger.

Erlichman, J. (1984). The pill-makers bluster, but is it just bluff? *The Guardian*, March 6th.

Fishlock, D. (1982). Pitfalls for biotechnology. In *The Business of Biotechnology*, pp. 69–86. London: Financial Times Business Information.

Hirose, Y. & Okada, H. (1979). Microbial production of amino acids. In *Microbial Technology*, 2nd edn, Vol. 1, ed. H. J. Peppler & D. Perlman, pp. 211–40. New York: Academic Press.

King, P. P. (1982). Biotechnology. An industrial view. *Journal of Chemical Technology and Biotechnology*, **32**, 2–8.

Klein, H. G. (1978). Antibiotics. In *Kirk–Othmer Encyclopaedia of Chemical Technology*, 3rd edn, Vol. 2, ed. H. F. Mark, D. F. Othmer, C. G. Overberger & G. T. Seaborg, pp. 809–19. New York: John Wiley & Sons.

289

Measday, W. (1977). The pharmaceutical industry. In *Structure of American Industry*, ed. W. Adams, pp. 250–84. New York: Macmillan.

Rapoport, C. (1983). The lop-sided world of the drug industry. *Financial Times*, 15 November.

Reuben, B. G. & Burstall, M. L. (1973). The new chemical industry from 1820 onwards. In *The Chemical Economy*, pp. 24–39. London: Longman.

Schumpeter, J. A. (1942). Monopolistic practices. In *Capitalism, Socialism and Democracy*, pp. 87–110. London: Unwin.

Tosaka, O., Enei, H. & Hirose, Y. (1983). The production of L-lysine by fermentation. *Trends in Biotechnology*, 1, (3), 70–73.

Walker, H. D. (1971). *Market Power and Price Levels in the Ethical Drug Industry*. Bloomington: Indiana University Press.

Watson, D. S. & Holman, M. A. (1977). General equilibrium and economic welfare. In *Price Theory and Its Uses*, 4th edn, pp. 418–37. Boston: Houghton Mifflin Co.

Chapter 3

Bach, N. G. (1960). How to get more accurate plant cost estimates. In *Cost Engineering in the Process Industries*, ed. C. H. Chilton, pp. 15–19. New York: McGraw Hill.

Crespi, R. S. (1982). *Patenting in the Biological Sciences*. Chichester: John Wiley & Sons.

Davies, W. (1967). *The Pharmaceutical Industry*. Oxford: Pergamon Press.

Demain, A. L. (1981). Industrial microbiology. *Science*, 214, 987–95.

Hand, W. E. (1958). Lang factors. *Petroleum Refiner*, 37, 331.

Kharbanda, O. P. & Stallworthy, E. A. (1982). Small is beautiful. In *How To Learn from Project Disasters: True Life Stories with a Moral for Management*, pp. 233–50. Aldershot: Gower.

Lang, H. J. (1948). Simplified approach to preliminary cost estimates. *Chemical Engineering*, June, 112–13.

Perry, R. (1984). What is a patentable biotechnological invention? In *The Proceedings of Biotech. '84 Europe*, pp. 45–56. Pinner: Outline Publications Ltd.

Scherer, F. M. (1970). *Industrial Market Structure and Economic Performance*, pp. 346–78. Chicago: Rand McNally.

Schumacher, E. F. (1976). *Small is Beautiful: Economics As If People Mattered*. New York: Harper & Row.

Schumpeter, J. A. (1934). The fundamental phenomenon of economic development. In *The Theory of Economic Development*, pp. 57–94. Oxford University Press.

Schumpeter, J. A. (1942). Monopolistic practices. In *Capitalism, Socialism and Democracy*, pp. 87–106. London: Unwin.

Scott, R. (1978). Working capital and its estimation for project evaluation. *Engineering and Process Economics*, 3, 105–14.

Vaughan, C. (1984). Systems for collaboration in advanced biotechnology. In *The Proceedings of Biotech. '84 Europe*, pp. 159–73. Pinner: Online Publications Ltd.

Wheaton, J. B. (1984). Licensing of biotechnology products. In *The Proceedings of Biotech. '84 Europe*, pp. 57–68. Pinner: Online Publications Ltd.

Chapter 4

Brown, O. M. R. (1983). Sucrose and molasses as feedstocks for fermentation processes. *Chemistry and Industry*, **3**, 95–7.

Hepner, L. & Associates. (1984). As quoted in *Chemical and Engineering News*, 19th March, 12.

Hill, L. D. & Mustard, A. (1981). Economic considerations in industrial utilization of cereals. In *Cereals: A Renewable Resource. Theory and Practice*, ed. Y. Pomeranz & L. Munck, pp. 25–54. St Paul Minn., USA: American Society of Cereal Chemists.

Phillips, J. A. & Humphrey, A. E. (1983). Process biotechnology for the conversion of biomass into liquid fuels. In *Liquid Fuel Developments*, ed. D. L. Wise, pp. 65–95. Boca Raton, Fla., USA: CRC Press.

Chapter 5

Atkinson, B. & Mavituna, F. (1983). *Biochemical Engineering and Biotechnology Handbook*, pp. 1033–6. London: Macmillan.

Bartholomew, W. H. & Reisman, H. B. (1979). Economics of fermentation processes. In *Microbial Technology*, 2nd edn, Vol. II, ed. H. J. Peppler & D. Perlman, pp. 463–96. New York: Academic Press.

Cooney, C. L. (1979). Conversion yields in penicillin production: theory *vs.* practice. *Process Biochemistry*, **14** (5), 31–3.

Coote, N. (1983). The impact of ethanol fuels on distillery design. In *Advances in Fermentation*, pp. 3–9. A supplement to *Process Biochemistry*. Rickmansworth, UK: Wheatland Journals.

Ericsson, M., Ebbinhaus, L. & Lindblom, M. (1981). Single cell protein from methanol: economic aspects of the Norprotein process. *Journal of Chemical Technology and Biotechnology*, **31**, 33–43.

Florent, J. & Ninet, L. (1979). Vitamin B_{12}. In *Microbial Technology*, 2nd edn, Vol. I, ed. H. J. Peppler & D. Perlman, pp. 497–519. New York: Academic Press.

Guidoboni, G. E. (1984). Continuous fermentation systems for alcohol production. *Enzyme Microbial Technology*, **6**, 194–200.

Keim, C. R. (1983). Technology and economics of fermentation alcohol – and update. *Enzyme and Microbial Technology*, **5**, 103–14.

Paturau, J. M. (1982). *By-products of the Cane Sugar Industry. An Introduction to their Industrial Utilization*, 2nd edn, pp. 279–85. Amsterdam: Elsevier.

Perlman, D. (1978). Vitamins. In *Economic Microbiology*, Vol. 2, *Primary Products of Metabolism*, ed. A. H. Rose, pp. 303–26. London: Academic Press.

Pirt, S. J. (1975). *Principles of Microbe and Cell Cultivation*, p. 64. Oxford: Blackwell.

Schierholt, J. (1977). Fermentation processes for the production of citric acid. *Process Biochemistry*, **12** (9), 20–21.

Skøt, G. (1983). Prospects of continuous culture in the production of enzymes. In *Advances in Fermentation*, pp. 154–9. A supplement to *Process Biochemistry*. Rickmansworth, UK: Wheatland Journals.

Sodeck, G., Modl, J., Kominek, J. & Salzbrunn, W. (1981). Production of citric acid according to the submerged fermentation process. *Process Biochemistry*, **16** (6), 9–11.

Chapter 6

Atkinson, B. & Mavituna, F. (1983). Downstream process engineeering. In *Biochemical Engineering and Biotechnology Handbook*, pp. 890–931. London: Macmillan.

Cooney, C. L. (1979). Conversion yields in penicillin production: Theory *vs.* practice. *Process Biochemistry*, **14** (5), 31–3.

de Flines, J. (1980). Biotechnology – its past, present and future. In *Biotechnology: a Hidden Past, a Shining Future*, pp. 12–17. Proceedings of the 13th International TNO Conference. The Hague: Netherlands Central Organisation for Applied Scientific Research TNO.

Fish, N. M. & Lilly, M. D. (1984). The interactions between fermentation and protein recovery. *Biotechnology*, **2**, 623–7.

Gow, J. S., Littlehailes, J. D., Smith, S. R. L. & Walter, R. B. (1975). SCP Production from methanol bacteria. In *Single Cell Protein II*, ed. S. R. Tannenbaum & D. I. C. Wang, pp. 370–84. Cambridge, Mass.: MIT Press.

Keim, C. R. (1983). Technology and economics of fermentation alcohol – an update. *Enzyme and Microbial Technology*, **5**, 103–14.

Kroner, K. H., Hustedt, H. & Kula, M. R. (1984). Extractive enzyme recovery: economic considerations. *Process Biochemistry*, **19** (5), 170–9.

Labuza, T. (1975). Cell collection: recovery and drying for SCP manufacture. In *Single Cell Protein II*, ed. S. R. Tannenbaum & D. I. C. Wang, pp. 69–104. Cambridge, Mass.: MIT Press.

Rosen, C. G. (1983). As quoted in *Biotechnology News*, **3**, No. 10, 5.

Chapter 7

Antrim, R. L., Colilla, W. & Schnyder, B. (1979). Glucose isomerase production of high fructose syrups. In *Applied Biochemistry and Bio-Engineeering, Vol. 2, Enzyme Technology*, ed. L. B. Wingard, E. Katchalski-Katzir & L. Goldstein, pp. 97–155. New York: Academic Press.

Bucke, C. (1983*a*). Immobilized cells. *Philosophical Transactions of the Royal Society, Series B*, **300**, 369–89.

Bucke, C. (1983*b*). Carbohydrate transformations by immobilized cells. *Biochemical Society Symposium*, **48**, 25–8.

Chibata, T., Tosa, T. & Sato, T. (1979). Use of immobilized cells systems to prepare fine chemicals. In *Microbial Technology*, 2nd edn, Vol. II, ed. H. J. Peppler & D. Perlman, pp. 433–61. New York: Academic Press.

Chibata, I., Tosa, T. & Takata, I. (1983). Continuous production of L-malic acid by immobilized cells. *Trends in Biotechnology*, **1** (1), 9–11.

Daniels, M. (1984). Immobilized enzymes: the economics of industrial operations. In *The Proceedings of Biotech. '84 Europe*, pp. 405–13. Pinner: Online Publications Ltd.

Denner, W. H. B., Reichelt, J. R. & Farrow, R. I. (1983). Legislation and regulation. In *Industrial Enzymology: The Application of Enzymes in Industry*, ed. T. Godfrey & J. Reichelt, pp. 111–37. London: Macmillan.

Dunnill, P. (1980). Immobilized cell and enzyme technology. *Philosophical Transactions of The Royal Society London, Series B*, **290**, 409–20.

Godfrey, T. (1983). Immobilized enzymes. In *Industrial Enzymology: The Applications of Enzymes in Industry*, ed. T. Godfrey & J. Reichelt, pp. 437–52. London: Macmillan.

Godfrey, T. & Reichelt, J. (1983). Introduction to industrial enzymology. In *Industrial Enzymology: The Applications of Enzymes in Industry*, ed. T. Godfrey & J. Reichelt, pp. 1–7. London: Macmillan.

Hemmingsen, S. H. (1979). Development of an immobilized glucose isomerase for industrial application. In *Applied Biochemistry and Bioengineering, Vol. 2. Enzyme Technology*, ed. L. B. Wingard, E. Katchalski-Katzir & L. Goldstein, pp. 157–83. New York: Academic Press.

Hepner, L. & Associates. (1983). As quoted in *Chemical and Engineering News*, September 12, 11–12.

Meers, J. L. (1983). In *Industrial and Diagnostic Enzymes. Philosophical Transactions of the Royal Society London, Series B*, **300**, 427.

Michaelis, L. & Ehrenreich, M. (1908). Die Adsorptimanalyse der Fermente. *Biochemische Zeitschrift*, **10**, 283–99.

Naher, G. (1983). In *Industrial and Diagnostic Enzymes. Philosophical Transactions of the Royal Society London, Series B*, **300**, 425.

Stenson, S. C. (1983). Catalysts: a chemical market poised for growth. *Chemical and Engineering News*, December 5th, 19–25.

Venkatasubramanian, K., Constantinides, A. & Vieth, W. R. (1978). Synthesis of organic acids and modification of steroids by immobilized whole microbial cells. In *Enzyme Engineeering, Vol. 3*, ed. E. K. Pye & J. J. Weetall, pp. 29–43. New York: Plenum Press.

Wandrey, C. & Flaschel, E. (1979). Process development and economic aspects of enzyme engineering. Acylase L-methionine system. *Advances in Biochemical Engineering*, **12**, 147–218.

Chapter 8

Abelson, P. H. (1983). As quoted by J. Haggin & J. H. Krieger, in Biomass becoming more important in US energy mix, *Chemical and Engineering News*, **61** (11), 28–30.

Adams, M. R. & Flynn, G. (1982). *Fermentation Ethanol: An Industrial Profile*. London: Tropical Development Research Institute.

Banks, F. E. (1980). *The Political Economy of Oil*. Lexington MA, USA: Lexington Books.

Coote, N. (1983). The impact of ethanol fuels in distillery design. In *Advances in Fermentation*, pp. 3–9. A supplement to *Process Biochemistry*. Rickmansworth, UK: Wheatland Journals.

Esser, L. & Schmidt, U. (1982). Alcohol production by biotechnology. *Process Biochemistry*, **17**, 46–9.

Flannery, R. J. & Steinschneider, A. (1983). Fermentation economics in relation to genetic engineeering. *Biotechnology*, **1**, 773–6.

Guidoboni, G. E. (1984). Engineeering for an economic fermentation. *Chemistry and Industry*, **12**, 439–43.

Keim, C. R. (1983). Technology and economics of fermentation alcohol – an update. *Enzyme and Microbial Technology*, **5**, 103–14.

Lantzke, U. (1982). Energy and the economy. In IEA/OECD World Energy Outlook, pp. 63–77. Paris: OECD.

Lyons, T. P. (1983). Alcohol Fermentation in the United States. In *Advances in Fermentation*, pp. 16–23. A supplement to *Process Biochemistry*. Rickmansworth, UK: Wheatland Journals.

Mauldin, G. L. & Phelan, P. F. (1978). *Gasohol.* (*A study by New Mexico Solar Institute.*) Las Cruces USA: New Mexico State University.

Mouris, B. (1984). Economics and energy balances of ethanol from sugar cane and sugar beet. *Chemistry and Industry*, **12**, 435–8.

Paturau, J. M. (1982). *By-products of the Cane Sugar Industry. An Introduction to their Industrial Utilization*, 2nd edn, pp. 194–270. Amsterdam: Elsevier.

Rothman, H., Greenshields, R. & Calle, F. R. (1983). *The Alcohol Economy: Fuel Ethanol and the Brazilian Experience*, pp. 89–104. London: F. Pinter.

Spivey, M. J. (1978). The acetone/butanol/ethanol fermentation. *Process Biochemistry*, **13** (11), 2–4, 25.

Stone, J. (1974). *Survey of Alcohol Fuel Technology*. An Interim Report to Mitre Corp., McLean, Va., USA.

Chapter 9

Greditor, A. S. (1982). *Nutritive Sweetener Update. Passing Through the Cyclical Trough.* New York: Drexel, Burnham, Lambert Inc.

Greditor, A. S. (1983). *Nutritive Sweetener Industry Outlook; The Expanding Importance of Ethanol.* New York: Drexel, Burnham, Lambert Inc.

Keim, C. R. (1983). Technology and economics of fermentation alcohol – an update. *Enzyme and Microbial Technology*, **5**, 103–14.

Chapter 10

Colleran, E., Barry, M., Wilkie, A. & Newell, P. J. (1982). Anaerobic digestion of agricultural wastes using the upflow anaerobic filter design. *Process Biochemistry*, **17**, 12–17.

Dunnill, P. (1981). Biotechnology and industry. *Chemistry and Industry*, **7**, 204–17.

Gibbons, W. R. & Westby, C. A. (1983). Fuel ethanol and high protein feed from corn and corn–whey mixtures in a farm-scale plant. *Biotechnology and Bioengineering*, **25**, 2127–48.

Hayes, T. D., Jewell, W. J., Dell'Ovoto, S., Fanfoni, K. J., Leuschner, A. P. & Sherman, D. F. (1980). In *Anaerobic Digestion*, ed. D. A. Stafford, B. I. Wheatley & D. E. Hughes, pp. 255–88. London: Applied Science Publishers.

Marchaim, U., Prochenski, D., Perach, Z., Lombrozo, E., Dvoskin, D. & Criden, J. (1982). The Israeli anaerobic digestion process – 'Nefah'. In *Anaerobic Digestion, 1981*, ed. D. E. Hughes *et al.*, pp. 203–16. Amsterdam: Elsevier.

Pitcher, W. H., Ford, J. R. & Weetall, H. H. (1976). The preparation, characterization and scale-up of a lactase system immobilized to inorganic supports for the hydrolysis of acid whey. *Methods in Enzymology*, **44**, ed. K. Mosbach, pp. 792–809. New York: Academic Press.

Prenosil, J. E., Stuker, E., Hediger, T. & Borne, J. R. (1984). Enzymatic whey hydrolysis in the pilot plant 'Lactohyd'. *Biotechnology*, **2** (5), 441–4.

Schellenbach, S. (1982). Case study of a farmer-owned and operated 1000 head feedlot anaerobic digester. In *Energy from Biomass and Wastes VI*, Chairman D. L. Klass, pp. 545–66. Chicago: Institute of Gas Technology.

Sitton, O. C., Foutch, G. L., Brook, N. L. & Gaddy, J. L. (1979). Ethanol from agricultural residues. *Process Biochemistry*, **14**, 7–10.

Szendry, L. M. (1983). The Bacardi Corporation digestion process for stabilizing rum distillery wastes and producing methane. In *Energy from Biomass and Wastes VII*, Chairman D. L. Klass, pp. 767–94. Chicago: Institute of Gas Technology.

Chapter 11

Anon. (1982). Index of biotechnology stocks. *Nature*, **299**, 101.

Anon. (1984*a*). Stock prices of selected biotechnology speciality firms. *Biotechnology*, **2** (**6**), 502.

Anon. (1984*b*). Monoclonal antibody firms face growing problems. *Biotechnology News*, **4** (**5**), 5.

Anon. (1984*c*). Lymphokine markets. *Genetic Technology News*, February, 6.

Cane, A. (1983). The tumours that save calves' lives. *Financial Times*, 10th October.

Curtin, M. E. (1983). Harvesting profitable products from plant tissue culture. *Biotechnology*, **1**, 649–57.

Dominguez, J. R. (1974). *Venture Capital*, pp. 1–10. Lexington MA, USA: Lexington Books.

Dyos, H. J. & Aldcroft, D. H. (1969). The creation of the railway system. In *British Transport: An Economic Survey from the Seventeenth Century to the Twentieth*, pp. 116–54. Leicester University Press.

Fishlock, D. (1982). Government support. In *The Business of Biotechnology*, pp. 27–47. London: Financial Times Business Information Ltd.

Hester, A. (1983). As quoted in B. A. Tokay. (1983).

Klausner, A. (1984). Microbial insect control using bugs to kill bugs. *Biotechnology*, **2**, 408–19.

Lee, V., Gurnsey, J. & Klausner, A. (1983). New trends in financing biotechnology. *Biotechnology*, **1**, 544–59.

Lisansky, S. G. (1984). As quoted by J. Erlichman in *The Guardian*, 9th October.

Murray, J. R. (1983). Patterns of investment in biotechnology. I. *Biotechnology*, **1**, 248–50.

Murray, J. R. (1984). Patterns of investment in biotechnology. II. *Biotechnology*, **2**, 332–3.

Netzer, W. J. (1984). Appropriate bean gene expression reported. *Biotechnology*, **2**, 596.

Powledge, T. M. (1984). Interferon on trial. *Biotechnology*, **2**, 214–28.

Tokay, B. A. (1983). The new biotechnology goes to market. *Chemical Marketing Reporter*, 19th September, 24–33.

Vaughan, C. (1982). As quoted in D. Fishlock. (1982), p. 52.

Wilson, T. (1984). Engineering tomorrow's vaccines. *Biotechnology*, **2**, 28–40.

Yachinski, S. (1983). Race for the power plants hots up. *Financial Times*, 10th October.

Chapter 12

Chattin, B. & Doering, O. C. (1984). An alcohol fuels programme; will it solve surplus grain production? *Food Policy*, **9**, 35–43.

van Duijn, J. J. (1984). Fluctuations in innovations over time. In *Long Waves in the World Economy*, ed. C. Freeman, pp. 19–30. London: Francis Pinter.

Freeman, C. (1984). *Long Waves in the World Economy*. London: Francis Pinter.

Kondratiev, N. D. (1935). The long waves in economic life. *Review of Economic Statistics, 1935*, 105–15.

Mowery, D. C. & Rosenberg, N. (1982). The influence of market demand upon innovation: a critical review of some recent empirical studies. In *Inside The Black Box: Technology and Economics*, ed. N. Rosenberg, pp. 193–241. Cambridge University Press.

Perlman, D. (1978). Stimulation of innovation in the fermentation industries, 1910–1980. *Process Biochemistry*, **13** (**5**), 3–5.

Rosenberg, N. (ed.) (1982*a*). *Inside The Black Box: Technology and Economics*. Cambridge University Press.

Rosenberg, N. (1982*b*). The historiography of technical progress. In *Inside The Black Box: Technology and Economics*, pp. 3–33. Cambridge University Press.

Schumpeter, J. A. (1939). *Business Cycles*. New York: McGraw Hill.

Wagner, C. K., Lipinsky, E. S., McClure, T. A. & Scantland, D. A. (1983). Integration of production of corn-derived fuels with animal feed production. In *Liquid Fuel Developments*, ed. D. L. Wise, pp. 189–204. Boca Raton, Fla., USA.: CRC Press.

Index